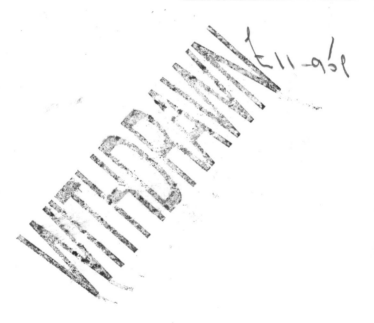

THE NEW GROVE
EARLY ROMANTIC MASTERS 2

THE NEW GROVE
DICTIONARY OF MUSIC AND MUSICIANS

Editor: Stanley Sadie

The Composer Biography Series

BACH FAMILY
BEETHOVEN
EARLY ROMANTIC MASTERS 1
EARLY ROMANTIC MASTERS 2
HANDEL
HAYDN
HIGH RENAISSANCE MASTERS
ITALIAN BAROQUE MASTERS
MASTERS OF ITALIAN OPERA
MODERN MASTERS
MOZART
SCHUBERT
SECOND VIENNESE SCHOOL
TURN OF THE CENTURY MASTERS
WAGNER

THE NEW GROVE®

Early Romantic Masters 2

WEBER   BERLIOZ
MENDELSSOHN

John Warrack
Hugh Macdonald
Karl-Heinz Köhler

M
MACMILLAN

Copyright © John Warrack, Hugh Macdonald, Karl-Heinz Köhler,
Eveline Bartlitz 1980, 1985
All other material © Macmillan Press Limited 1985

First published in
The New Grove Dictionary of Music and Musicians®,
edited by Stanley Sadie, 1980

The New Grove and *The New Grove Dictionary of Music and Musicians*
are registered trademarks of Macmillian Publishers Limited, London

First published in UK in paperback with additions 1985 by
PAPERMAC
a division of Macmillan Publishers Limited
London and Basingstoke

First published in UK in hardback with additions 1985 by
MACMILLAN LONDON LIMITED
4 Little Essex Street London WC2R 3LF
and Basingstoke

British Library Cataloguing in Publication Data
Warrack, John
Early romantic masters 2.—(The composer
biography series)
1. Music—Europe—History and criticism
2. Composers
I. Title II. Macdonald, Hugh, *1940–*
III. Köhler, Karl-Heinz IV. The New Grove
dictionary of music and musicians V. Series
780′.92′2   ML240

ISBN 0–333–39013–X (hardback)
ISBN 0–333–39014–8 (paperback)

First American edition in book form with additions 1985 by
W. W. NORTON & COMPANY
New York and London

ISBN 0-393-01692-7 (hardback)
ISBN 0-393-30096-X (paperback)

Printed in Great Britain

Contents

List of illustrations

Cover: Wolf's Glen scene in Weber's 'Der Freischütz': coloured aquatint by C. Lieber, with figures by Carl August Schwerdgeburth, of Carl Wilhelm Holdermann's design for the 1822 Weimar production (Staatliche Kunstsammlungen, Weimar).

Illustration acknowledgments

We are grateful to the following for permission to reproduce illustrative material: Royal College of Music, London (figs.1, 4); Sächsische Landesbibliothek, Dresden (fig.2); Royal Opera House, Covent Garden, London (fig.3); Deutsche Staatsbibliothek, Musikabteilung, Berlin (figs.5, 7); Deutsche Fotothek, Dresden (fig.6); Académie de France, Rome (fig.8); Richard Macnutt, Tunbridge Wells (figs.11, 12, 18, 23); Bibliothèque Nationale, Paris (figs.13, 14); Bodleian Library, Oxford (fig.17); Mansell Collection, London (fig.19); Staatsbibliothek Preussischer Kulturbesitz, Berlin (fig.20); Central Libraries, Birmingham (fig.21); Archiv für Kunst und Geschichte, Berlin (fig.22); Music Division, Library of Congress, Washington, DC (fig.25); Staatliche Kunstsammlungen, Weimar (cover)

General abbreviations

A	alto, contralto [voice]	κ	Köchel catalogue [Mozart; no. after / is from 6th edn.]
addns	additions		
Anh.	Anhang [appendix]		
aut.	autumn		
		lib	libretto
B	bass [voice]		
b	bass [instrument]	Mez	mezzo-soprano
Bar	baritone [voice]	movt	movement
bc	basso continuo		
bn	bassoon	n.d.	no date of publication
BWV	Bach-Werke-Verzeichnis [Schmieder, catalogue of J. S. Bach's works]		
		ob	oboe
		orch	orchestra, orchestral
		orchd	orchestrated (by)
c	circa [about]	org	organ
cl	clarinet	ov.	overture
collab.	in collaboration with		
Cz.	Czech	perc	percussion
		perf.	performance, performed (by)
db	double bass		
ded.	dedication, dedicated to	prol	prologue
		pubd	published
edn.	edition		
		qnt	quintet
facs.	facsimile	qt	quartet
fl	flute		
frag.	fragment	R	photographic reprint
		recit	recitative
gui	guitar	repr.	reprinted
		rev.	revision, revised (by/for)
hn	horn		
		S	soprano [voice]
inc.	incomplete	str	string(s)
inst	instrument, instrumental		
		T	tenor [voice]
J	Jähns catalogue [Weber]	timp	timpani
Jb	Jahrbuch [yearbook]	transcr.	transcription, transcribed (by/for)
Jg.	Jahrgang [year of publication/volume]		
		trbn	trombone

U.	University	vc	cello
		vn	violin
v, vv	voice, voices		
va	viola	ww	woodwind

Symbols for the library sources of works, printed in *italic*, correspond to those used in *Répertoire International des Sources Musicales*, Ser. A.

Bibliographical abbreviations

AcM	*Acta musicologica*
ADB	*Allgemeine deutsche Biographie* (Leipzig, 1875–1912)
AMw	*Archiv für Musikwissenschaft*
AMZ	*Allgemeine musikalische Zeitung*
AMz	*Allgemeine Musik-Zeitung*
BMw	*Beiträge zur Musikwissenschaft*
BWQ	*Brass and Woodwind Quarterly*
DJbM	*Deutsches Jahrbuch der Musikwissenschaft*
FAM	*Fontes artis musicae*
Grove 1 (–4)	G. Grove, ed.: *A Dictionary of Music and Musicians*, 2nd–4th edns. as *Grove's Dictionary of Music and Musicians*
HJb	*Händel-Jahrbuch*
IRASM	*International Review of the Aesthetics and Sociology of Music*
JAMS	*Journal of the American Musicological Society*
JbMP	*Jahrbuch der Musikbibliothek Peters*
ML	*Music and Letters*
MMR	*The Monthly Musical Record*
MQ	*The Musical Quarterly*
MR	*The Music Review*
MT	*The Musical Times*
NBJb	*Neues Beethoven-Jahrbuch*
NZM	*Neue Zeitschrift für Musik*
ÖMz	*Österreichische Musikzeitschrift*
PMA	*Proceedings of the Musical Association*
PRMA	*Proceedings of the Royal Musical Association*
RBM	*Revue belge de musicologie*
RdM	*Revue de musicologie*
ReM	*La revue musicale*

RHCM	*Revue d'histoire et de critique musicales* [1901]; *La revue musicale* [1902–10]
RMG	*Russkaya muzïkal'naya gazeta*
RMI	*Rivista musicale italiana*
SIMG	*Sammelbände der Internationalen Musik-Gesellschaft*
VMw	*Vierteljahrsschrift für Musikwissenschaft*
ZfM	*Zeitschrift für Musik*
ZIMG	*Zeitschrift der Internationalen Musik-Gesellschaft*
ZMw	*Zeitschrift für Musikwissenschaft*

Preface

This volume is one of a series of short biographies derived from *The New Grove Dictionary of Music and Musicians* (London, 1980). In its original form, the text was written in the mid-1970s, and finalized at the end of that decade. For this reprint, the texts have been re-read and modified, those for Weber and Berlioz by their respective authors and that for Mendelssohn by R. Larry Todd, who has also made substantial changes in the light of his research to the Mendelssohn work-list (originally drawn up by Eveline Bartlitz) and has supplemented the bibliography.

The fact that the texts of the books in this series originated as dictionary articles inevitably gives them a character somewhat different from that of books conceived as such. They are designed, first of all, to accommodate a very great deal of information in a manner that makes reference quick and easy. Their first concern is with fact rather than opinion, and this leads to a larger than usual proportion of the texts being devoted to biography than to critical discussion. The nature of a reference work gives it a particular obligation to convey received knowledge and to treat of composers' lives and works in an encyclopedic fashion, with proper acknowledgment of sources and due care to reflect different standpoints, rather than to embody imaginative or speculative writing about a composer's character or his music. It is hoped that the comprehensive work-lists and extended bibliographies, indicative of the origins of the books in a reference work, will be valuable to the reader who is eager for full and accurate reference information and who may not have ready access to *The New Grove Dictionary* or who may prefer to have it in this more compact form.

S.S.

CARL MARIA VON WEBER

John Warrack

CHAPTER ONE

Life

I Youth and first appointments

All his life Carl Maria von Weber believed that he could trace his ancestry back to one Johann Baptist Weber, who had been ennobled by Ferdinand II in 1622; however, the earliest known member of the family was a miller named Hans Georg Weber who died in 1704, and the baronial 'von' does not appear until the composer's father Franz Anton Weber seems quietly to have appropriated it. Franz Anton's second wife was Genovefa Brenner (1764–98); her experience as a singer and actress was useful to him when he formed his Weber Theatre Company. From 1779 he had been Kapellmeister to the Prince-Bishop of Lübeck at Eutin; and there Carl Maria was born. For long the date was accepted, from Weber's *Autobiographische Skizze*, as 18 December 1786. However, the baptismal date in the Landeskirche for Carl Maria Friedrich Ernst von Weber is registered as 20 November, suggesting either the 18th or 19th as the birthday: Weber later accepted the 18th, with the more enthusiasm as it was also the birthday of his wife Caroline.

A weakly child, with a damaged right hip-bone that gave him a permanent limp, Weber is said to have first learnt to play among the scenery of his father's travelling company. His first music lessons, not a success, came from his half-brother Fridolin. When the company

1

paused in Hildburghausen, he was able to have more systematic lessons from Johann Peter Heuschkel, and from then his development as a musician was rapid. He studied further under Michael Haydn in Salzburg, and from this period dates his first work, a set of six fughettas J1–6; these were generously received by Rochlitz in the *Allgemeine musikalische Zeitung*. In Munich, he studied under Johann Nepomuk Kalcher, writing among other works his first opera, *Die Macht der Liebe und des Weins* (lost), and the first draft of his so-called *Grosse Jugendmesse*; at this time he also acquired a knowledge of Singspiels and French operas at the Munich theatre. Hoping to facilitate the printing of his music, he worked briefly under the inventor of lithography, Aloys Senefelder, but the drudgery attached to commercial lithography soon impelled him and his father to give up the lithography business they had tentatively established in Freiberg. It was in Freiberg that Weber wrote his second opera, *Das Waldmädchen*, to a text by the director of a travelling opera company, Carl von Steinsberg, who performed it on 24 November 1800. The music, which occasioned a public controversy between Franz Anton and the conductor, is now lost apart from two numbers.

Returning through Chemnitz and Munich, Weber reached Salzburg again in November 1801. There, under Michael Haydn, he revised his mass and completed his third opera, *Peter Schmoll und seine Nachbarn* (J8), which Haydn declared was composed 'with much fire and delicacy, and appropriately to the text'. Plans for a production in Augsburg, where Weber's half-brother Edmund was working, were postponed; and only after further travels, including to

2

Hamburg and Eutin, and the composition of a number of minor piano pieces and songs for concerts, did the Webers return to Augsburg for the production of *Peter Schmoll*, probably in March 1803.

From Augsburg, Weber moved to Vienna with the intention of studying under Joseph Haydn; however, he immediately fell under the influence of the Abbé Vogler, whose pupil he became and for whom he prepared the vocal score of the opera *Samori* (J39). He also wrote two sets of piano variations (the second with violin and cello parts ad lib) on themes from Vogler's operas (J40 and 43). He composed little else during nearly nine months in Vienna, preferring to study hard, to absorb Romantic impressions (including a fascination with folksong) and to indulge his talent for singing to his own guitar accompaniment in the taverns of the city and its surroundings. He in turn impressed Vogler enough to earn a recommendation, together with his new friend Johann Gänsbacher, for the post of Kapellmeister at Breslau. When Gänsbacher declined the nomination, Weber accepted, and, having collected his father again in Salzburg, he arrived in Breslau on 11 June 1804.

Despite the support of the theatre's Dramaturg, Johann Gottlieb Rhode, Weber immediately encountered opposition. Not yet 18, he was resented by the leader of the orchestra, who coveted the post, and he antagonized others inside and outside the theatre by the searching nature of his reforms: these included the re-arrangement of the orchestra's seating plan, a more demanding rehearsal schedule, the revision of the repertory to exclude feeble but popular works, and the super-annuation of elderly singers. However, he was supported by several prominent musicians in the city, especially the organist Friedrich Wilhelm Berner, and his performances won respect. He also wrote, to a text by Rhode, some numbers of a projected opera, *Rübezahl*: three numbers survive (J44–6). The end of this appointment came through an accident. Arriving one night to go over the *Rübezahl* music, Berner found Weber prostrate: he had absent-mindedly drunk from a wine bottle which Franz Anton had filled with engraving acid. He was ill for two months, and never recovered his singing voice; and when he returned to work, he found that his enemies had undone all his reforms. He resigned.

Through a pupil, he secured in the autumn of 1806 an appointment at Carlsruhe, Upper Silesia (now Pokój, Poland) as Intendant to Duke Eugen of Württemberg-Öls, who had established there a kind of miniature Romantic Versailles. The duke himself was a keen musician who played the oboe; in this period, which he later described as 'a golden dream', Weber was encouraged to resume the composing he had been obliged to neglect in Breslau. For the ducal orchestra he wrote two symphonies (J50 and 51), which omit clarinets and have prominent oboe parts; he also wrote some sets of variations and the first version of his Horn Concertino (J188). Leaving Carlsruhe, for reasons that are obscure, he returned briefly to Breslau, leaving again hurriedly on being recognized by a creditor, and after a short concert tour arrived in Stuttgart on 17 July 1807 to take up a new appointment as secretary to the duke's brother Ludwig.

II From Stuttgart to Prague

Württemberg, constituted a kingdom by Napoleon, was ruled over by Ludwig's elder brother Friedrich, a corrupt despot. A mutual dislike quickly arose between him and Weber; Weber's necessary contacts with him, often as the bearer of begging notes from his employer, led to difficulties, and on one occasion to a brief imprisonment in a room of the castle for a practical joke. Although he enjoyed some aspects of the backward feudal life of Stuttgart, including the chance of entertaining the younger court nobles with his piano and guitar improvisations, he found greater stimulus in the company of poets and intellectuals in the city. These included F. C. Hiemer, who reworked the *Waldmädchen* libretto as a new text for the opera *Silvana* (J87), and the composer Franz Danzi, who, as an elder friend and adviser, encouraged him to resume hard work on composition. Stimulus of a different kind came from an affair with one of the opera sopranos, Gretchen Lang, for whom he wrote a number of songs (16 in all date from Weber's Stuttgart period). There was also the encouragement to write some piano duets (J81–6) out of music lessons given to members of the duke's family; but for himself, already established as a virtuoso pianist, he wrote only four works, a set of variations (J55), the *Grande polonaise* (J59), the *Momento capriccioso* (J56) and three movements to complete his Piano Quartet (J76). Another piece, a set of variations for piano and violin (J61), was highly praised by Rochlitz, who also provided him with the text for the cantata *Der erste Ton* (J58). He also wrote, for Danzi, the incidental music for a production of Schiller's *Turandot* (J75).

In April 1809, Franz Anton arrived without warning. Now past 75, proud of his son but unpredictable in his behaviour, he set about meddling in affairs. A crisis came when he used some money, entrusted to Weber by Duke Ludwig for the purchase of horses, to settle his

5

own debts. Weber attempted to rescue the situation by borrowing, but on being sued for the recovery of the loan was arrested and imprisoned in an inn. Various charges brought against him were disposed of, but at the price of banishment from Württemberg. On 26 February 1810 the Webers were awakened by a police officer and escorted to the frontier.

Having settled his father in Mannheim, in the house of his namesake (but not relative) Gottfried Weber, and given a concert, Weber moved on to Heidelberg. There he was welcomed by the cellist Alexander von Dusch, and thanks to letters of introduction from Danzi was making good progress in the town until involvement in a student disturbance compelled him to return to Mannheim. He gave concerts to raise money to settle the Stuttgart debts, including one with the successful première of *Der erste Ton*, but wrote little apart from beginning his First Piano Concerto (J98); he also spent much time with Gottfried Weber and Dusch in discussion, mutual criticism and the singing and playing of folksongs. However, he soon felt the need to move to Darmstadt, where Vogler was living and teaching a group of pupils who included not only Gänsbacher but Jacob Beer (Meyerbeer). Vogler arranged various concerts for him, and some which they undertook together, as well as continuing his instruction.

Although this was not a very fertile period for composition, it was an important one for Weber's future development. He was 23; though small, slight and lame, unprepossessing in appearance, he had a liveliness and charm of manner that impressed his friends, and a fluency and articulateness that tended to make him dominate the company. He was delicately recep-

tive to Romantic impressions, literary and pictorial
as well as musical, and at this time invented from
various tunes, absorbed on his wanderings, themes that
were to recur later in his music. He was also fascinated
by his discovery of the story of *Der Freischütz*, for
which he immediately planned an opera with Dusch as
librettist; but the pressure of business kept Dusch from
completing the work, thereby postponing the project
until Weber's powers were equal to the task. In
Darmstadt he also wrote criticism, founded a
Harmonischer Verein for the propagation of the music
of its members (including Dusch, Meyerbeer, Gottfried
Weber, Gänsbacher, Danzi, Berner and himself) and
began work on a novel, never to be completed,
Tonkünstlers Leben. His compositions included the
completion of the First Piano Concerto and the *Six
sonates progressives* for violin and piano (J99–104); he
also began work on the Singspiel *Abu Hassan* (J106).

On 14 February 1811 Weber left Darmstadt on a
concert tour, including on his itinerary Bamberg (where
he first met E. T. A. Hoffmann), and reaching Munich
on 14 March. There he was given the chance of a
concert, which he shared with a friend he had met in
Darmstadt, the clarinettist Heinrich Baermann: among
his works, the new Clarinet Concertino (J109) was
especially well received, so much so that the king immediately commissioned two full-scale clarinet concertos (J114 and 118), and the orchestra besieged
Weber for other concertos, of which he wrote only one
for bassoon (J127). The success of *Abu Hassan* on 4
June convinced him anew of his calling as an opera
composer. Despite the success of his music in Munich,
feeling intoxicated rather than nourished by the social

round and unconsoled for the loss of his Mannheim friends by a succession of love affairs, the restless composer turned down an offer of the post of Kapellmeister at Wiesbaden and set off for Switzerland.

Recognized while crossing Württemberg territory, he was arrested and reimprisoned, but shortly released and sent on his way. He gave concerts in several Swiss towns, and planned a *Musikalische Topographie Deutschlands* that was to be a guide-book for travelling musicians. He also spent much time in walking, absorbing impressions of nature which he communicated to his friends in letters, before returning to Munich. Rapidly growing dissatisfied with his life there again, he planned a concert tour with Baermann, and they set off on 1 December. A few days later, they reached Prague, where their concerts were successful, and Weber made contact with the opera director Johann Liebich. They travelled on by way of Dresden and Leipzig to Gotha, where the enthusiastic but eccentric duke demanded a great deal from Weber, then to Weimar, where he had a stiff meeting with Goethe, a cordial one with Wieland, and a fruitful one with Pius Alexander Wolff, the author of *Preciosa*. On 20 February they arrived in Berlin.

In the competitive intellectual atmosphere of Berlin, Weber found a sharp contrast from the backward feudal states to which he was accustomed. He managed to have *Silvana* produced at the court theatre, and on 10 July it had a successful performance; for the *Liedertafel* he wrote a number of male-voice choruses that reflect the patriotism and national aspiration being openly expressed in the city. He also made valuable new friends, especially Hinrich Lichtenstein, a zoology professor

8

1. Carl Maria von Weber: portrait by John Cawse

who quickly recognized the novel emotional flavour of Weber's music and its relevance to the atmosphere of emergent nationalism, and Count Carl von Brühl, the Intendant of the opera. His music of this period includes the First Piano Sonata (J138).

Aware of the value of his stay in Berlin, yet also seeking greater stability in his career, Weber left at the end of August, visiting Gotha and Weimar. In October he resumed work on his Second Piano Concerto (J155), whose Rondo he had written the previous autumn, and played it in Gotha on 17 December. By way of Weimar and Leipzig, he reached Prague on 12 January 1813.

III Prague

Weber was met in Prague with the news that Wenzel Müller had resigned from the opera, and that he was being offered the post. At first reluctant to accept, he was persuaded partly by the enthusiasm of Liebich and other members of the operatic establishment, also by the chance of finally discharging his Stuttgart debts. Further, there was the challenge to make something of an opera house that had suffered a decline in the war, and to do so, on a three-year contract, in more stable conditions than those of his present life.

So busily did he throw himself into his work at the theatre that he had little time for original composition during the rest of 1813. Intending to build up the German company, he made a recruiting trip to Vienna (being anticipated by a letter from his future wife, Caroline Brandt, requesting a post). In Vienna, he held auditions, engaging among other artists the violinist Franz Clement (dedicatee of Beethoven's Violin Concerto) as his leader, and he also gave a concert himself. Back in Prague, having recovered from an illness during which he began to suffer chest pains, he set about reorganizing the entire system at the theatre. Characteristically, his methodical and all-embracing approach to opera presentation extended to conditions and details of employment for every member of the staff, and to recataloguing the library, as well as revising rehearsal schedules, holding discussions with scene-painters and wardrobe mistresses, and beginning to form of his new singers the nucleus of a company. There were naturally resentments; and when he found these expressed in his presence in Czech, he set himself to learn the language. The controversy he caused was not calmed by a stormy affair with a soubrette, Therese Brunetti, who treated him abominably, but from whom, in his lonely and sick state, he found it difficult to part.

On 12 August the new company began rehearsals with Spontini's *Fernand Cortez*. In all, as his notebook and register records, Weber was to give 62 operas by more than half as many composers during his Prague years. The first ones were almost exclusively French, chiefly by Spontini, Cherubini, Isouard, Boieldieu and Dalayrac (later also by Grétry, Catel and Méhul), for he recognized that it was here, rather than in the limited history of the German musical stage, that the example for Romantic opera was to be found. Scarcely an opera in his entire repertory is not in some way relevant to his intentions for the development of German Romantic opera; and in many of them there were specific examples for his own operas.

On 11 December 1813, Caroline Brandt arrived, making her début in the title role of Isouard's *Cendrillon* on New Year's Day. It was a part that suited her soubrette talent well, and it won her popularity in Prague and the jealousy of Therese Brunetti, from whom Weber with difficulty disentangled himself. For his benefit performance on 15 January he staged *Don Giovanni*, with Caroline as Zerlina; his solicitude over a slight stage accident she suffered drew them closer together. In July, exhausted by the strain of his first season, he left for a spa to recover his precarious health; at the end of the month he went to Berlin, where he took an enthusiastic part in the victory celebrations and contemplated, among several possible opera projects, *Tannhäuser* to a text by Brentano. After an unsuccessful production of *Silvana* (5 September), he left for Gotha, where he wrote some of the *Leyer und Schwert* songs, reluctantly returning to Prague before his leave had expired under pressure from Liebich to rescue the company's deteriorating standards. His sense of unrest was increased by an offer from the Berlin opera (a post he seriously considered, but which went in the end to Andreas Romberg) and by Caroline's hesitation about accepting his proposals of marriage. He was also bitterly disappointed when the public he had been nursing, and had begun to respect, failed to appreciate his production of *Fidelio* in November. However, the next March he had a financial success with his benefit performance of *Così fan tutte*, and among much other activity he wrote some pieces for visiting virtuosos, the clarinettist Hermstedt and the flautist father and son Caspar and Anton Fürstenau, and also the finest of his sets of piano variations, on a Russian song, *Schöne Minka* (J179).

Further upsets, including the sudden arrival of his difficult half-brother Fridolin and the renewed jealousy of Caroline, led Weber to anticipate his annual leave and set off for Munich on 6 June 1815. He had parted from Caroline on poor terms, but soon they were writing warmly to each other; Weber's letters reveal a depth of despair about his talent and his purpose as an artist which he did not uncover to any other correspondent. Going to church for the thanksgiving service for Waterloo, he encountered Wohlbrück and together they planned the cantata *Kampf und Sieg* (J190). Although very successful with his concerts and very popular with his friends, he was unhappy at what appears to have been a break with Caroline (her letters are lost). He began writing the *Grand duo concertant* (J204), and for Rauch revised his Horn Concertino (J188). By early September he was back in Prague.

Despite the anxieties about Caroline, Weber was relieved to return, and excited to find his enthusiasm for composition, and hence his faith in himself, rekindled by *Kampf und Sieg*. Only a few unimportant compositions, chiefly songs for interpolation in stage pieces, interrupted his work on the cantata. He also hit upon the idea of winning the Prague public round to his repertory by writing introductory articles in the newspapers: the first of these essays, on Meyerbeer's *Wirt und Gast*, appeared in October, and in them some of his most important observations on opera are to be found. He also wrote a very full introduction to *Kampf und Sieg*, in order to alert attention to its mixture of narrative, representation and meditation in quasi-dramatic terms. The work was well received by a small audience on 22 December (a storm kept many at home). Neither this success, nor reconciliation with Caroline in the autumn, weakened his resolve to leave Prague. Still less did a letter from the opera directors to Liebich complaining about the decline of the opera, to which Weber replied defending himself articulately and with dignity, emphasizing the importance of ensemble and of developing a German tradition in opera. Suffering from his lame hip and from sore throats in the damp Prague winter, he was taken in as a lodger by Caroline's mother. At Easter he sent in his resignation, with effect from the autumn.

On 5 June 1816, Weber left for Berlin again, passing through Dresden, where he met the royal equerry Count Heinrich Vitzthum and was presented with a snuffbox in acknowledgment of the score of *Kampf und Sieg*. His Berlin concerts were poorly attended; leaving on 9 July, he visited Leipzig, where he declined the offer of the opera on the grounds that he did not wish to commit himself to private enterprises again, and travelled on to Carlsbad, perhaps for the cure but also to renew discussions with Vitzthum, newly appointed Intendant at Dresden. He returned to Prague on 18 July to find Liebich bedridden and the affairs of the opera once more in disarray. He did his best to clear them up and document matters for his successor, and then sent Liebich his final resignation. The company saw him off to Berlin, together with Caroline and her mother, on 7 October.

It was the welcome Weber was given in Berlin that helped to remove the last of Caroline's doubts about accepting him; among her reasons for hesitating had been commitment to her career, which was not developing very fast, and doubts about his, which clearly was. He announced his betrothal on the day of a total eclipse of the sun, which re-emerged to illuminate the scene with a piece of dramatic timing perhaps engineered by

Weber and certainly not lost on him. When Caroline left on 20 November to fulfil engagements in Dresden, he settled contentedly to work at composition. Many of his best songs date from this period, including one of the first groups that can properly be called a cycle, *Die Temperamente beim Verluste der Geliebten* (J200–03). He also completed the *Grand duo concertant* and two more piano sonatas, no.2 in A♭ (J199) and no.3 in D minor (J206). Negotiations with Vitzthum had been proceeding throughout the autumn, and on Christmas morning he received the official letter appointing him Royal Saxon Kapellmeister.

IV **Dresden**

The city's operatic traditions were Italian, with *opera seria* that were supported by the court, and Italian singers and composers held the principal posts at the court opera. It was largely in order to give Saxony an honourable artistic position in the new Germany, when it had suffered political and military disgrace in the War, that Vitzthum wished to develop a German opera; he turned to Weber as the best man for the purpose. With difficulty, he negotiated the appointment of Weber as music director; and on 13 January 1817 Weber arrived in Dresden, unaware that this title ranked below Royal Kapellmeister, a post held by Morlacchi. When, a few days later, he learnt that this was to rate himself and German opera below Morlacchi and Italian opera, he resigned; but this move allowed the skilful Vitzthum to secure the two titles as equal ranking in court status. It could not, of course, still the rivalry with Morlacchi, which continued with varying degrees of bitterness throughout Weber's life.

Weber set about building up the German Opera with his usual thoroughness in every theatrical department. The resources were poor: for instance, the chorus upon which he placed considerable importance was thin in numbers (still only 32 by September) and ill-trained dramatically. His reforms extended to scenery and lighting, and later to the introduction of a new system of rehearsal schedules that included readings of the text: as in Prague, his intention was to work towards the more unified work of art which he saw as the peculiarly German contribution to opera. The repertory was also orientated, as in Prague, towards the development of Romantic opera. On 30 January he opened with Méhul's *Joseph*, following this with two less successful productions before visiting Prague to recruit from the now insolvent company there. Difficulties in establishing a *modus vivendi* with the Italian faction in Dresden multiplied so much that he even accepted an offer from Berlin; when this fell through (after the burning down of the Berlin theatre) Vitzthum used the occasion to improve Weber's position in Dresden with a permanent appointment. In one of the literary gatherings that were a feature of Dresden life, Weber met Friedrich Kind, with whom he revived the idea of an opera on *Der Freischütz*; but he had little time for composition apart from official commissions (including music for a festival play, *Der Weinberg an der Elbe* (J222), in which the four-year-old Wagner appeared as an angel). On 4 November 1817, in Prague, he married Caroline.

Required to engage singers for the German Opera, Weber and his bride spent their honeymoon travelling; to eke out their expenses they gave concerts. On his return Weber renewed his efforts, against opposition

16

from the court and from the Italians, to improve the conditions for giving opera, eventually achieving his way over the re-seating of the orchestra. He also wrote his *Missa sancta* in E♭ (J224) and began work on *Der Freischütz* (J277). Happily established with Caroline, he took a summer residence outside the city at Hosterwitz, where, among many other works, he composed some occasional music commissioned by the king. He returned to the city at the end of August 1818, exhausted by his summer's work and with the symptoms of his tuberculosis asserting themselves. The rest of the year was occupied in work in the theatre and in the composition of the *Missa sancta* in G (J251). On 22 December, a daughter was born, only to die at the end of March, when Weber himself was seriously ill.

Commissioned to write an opera, Weber had planned *Alcindor* with Kind; but when this was cancelled at the end of June 1819 he turned to instrumental pieces. To the summer belong the *Rondo brillante* (J252), the Trio for flute, cello and piano (J259), the *Aufforderung zum Tanze* (J260) and the *Polacca brillante* (J268), as well as two movements of the Fourth Piano Sonata and parts of *Der Freischütz*. His visitors included Marschner at Hosterwitz and Spohr in Dresden; and by 6 December he was able to write to Brühl in Berlin that the new opera would be ready in March to open the new Schauspielhaus. Relations improved with Morlacchi, only to deteriorate after a tactless article of Weber's deploring Meyerbeer's italianization. On 13 May 1820 *Der Freischütz* was finished, except for Aennchen's second aria, added the following year.

Before completing the opera Weber already had plans for incidental music to Wolff's *Preciosa* and a comic opera, *Die drei Pintos*, to a libretto by Theodor Hell; and to find more convenient circumstances in which to compose them, he took a house at Cosels Garten, nearer the town than Hosterwitz. The *Preciosa* music (J279) was finished on 15 July; and *Die drei Pintos* (JAnh.5) was begun. With news from Berlin of the postponement of *Der Freischütz*, he made a concert tour of north Germany, stopping in Leipzig, Halle, Göttingen, Hanover, Bremen and other towns. After visiting Eutin, he travelled on to Kiel and made the crossing to Denmark, where his successful concerts included the first hearing of the *Freischütz* overture. Back in Dresden on 4 November, he devoted the rest of the year largely to the theatre and to renewed work on his novel *Tonkünstlers Leben*.

In the New Year, 1821, he took on a pupil, Julius Benedict. On 14 March, *Preciosa* with Weber's music had a successful Berlin première; and early in May, beginning two months leave made possible by the redecoration of the theatre, he and Caroline set off for Berlin, arriving at the Beers' house on the 4th. He took with him sketches of an F minor piano concerto that was eventually to be the *Konzertstück* (J282), finished on the day of the *Freischütz* première, 18 June.

In Berlin, as in Dresden, Weber found himself elected the champion of the cause of German opera. Here, however, the other side was represented by the elaborate court opera of Friedrich Wilhelm III as upheld by the imposing spectacle operas of Spontini. The première of *Olympie* on 14 March had been merely brilliant; that of *Der Freischütz* was a triumph, in Berlin and soon throughout Germany, for in this new manner of opera based on German folklore and country life, and in its idiom close to the contours of German folksong, the nation sensed that it had found its musical voice. Although *Der Freischütz* was not without its adversaries, including those such as Spohr with a commitment to German opera, it became immediately and enduringly popular throughout Europe: no other German opera has ever been taken up so widely and so rapidly.

Although anxious not to exacerbate the situation between himself and Spontini, one not of his own making, Weber found himself treated like a conquering general; as Spitta wrote (*Grove 1–4*), 'The immediate success of Weber's work may not have more than equalled that of *Olympie*, but it soon became evident that the chief effect of the latter was astonishment, while the former set the pulse of the German people beating'. On 30 June, with success ringing in his ears, Weber returned to Dresden, having clinched his Berlin triumph with the *Konzertstück*. But tales from Berlin did not smooth his path in Dresden, where he remained subject to the usual restrictions and irritations. *Der Freischütz* was kept off the stage by intrigues; and when he was refused permission to produce *Die drei Pintos*, Weber had his first haemorrhage.

On 11 November he received a letter from Barbaia, lessee of the Vienna Kärntnertortheater, requesting an opera for the 1822–3 season. The proposal was for an opera 'in the style of *Der Freischütz*'; but determined to answer his critics with a grand opera, and one that would be an advance rather than a repetition, he turned to Helmina von Chezy, a minor Romantic poet living in Dresden, and accepted her suggestion of *Euryanthe*. Work on *Pintos* was again postponed, as *Der Freischütz*

19

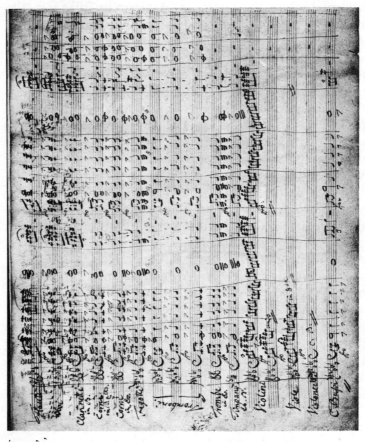

2. Autograph MS of the overture to Weber's 'Euryanthe', composed 1822–3

20

was now due in the New Year and one of Weber's
operas was thought to be enough for Dresden; he spent
the rest of the year looking after Morlacchi's commit-
ments as well as his own, working on a scheme for
subscription tickets, and nursing Caroline (who had
miscarried in 1820) through her latest pregnancy.

The Dresden production of *Der Freischütz* on 26
January 1822 was a success; a fortnight later Weber left
for Vienna by way of Prague (where he heard Henriette
Sontag): though depressed by the Vienna production
of *Der Freischütz*, he admired Wilhelmine Schröder's
Agathe. He met Schubert, and tried to heal a breach
with Salieri; he was ill, and though he did conduct *Der
Freischütz* at Schröder's pleading, the very success
made him fear for *Euryanthe* by comparison. He was
back in Dresden on 26 February, when he was obliged
to resume work for the still absent Morlacchi as well as
in his own company, and also to move house from the
Altmarkt to the Galleriestrasse. On 25 April his first
son (and future biographer) Max Maria was born.

By 15 May Weber was able to move his family to
Hosterwitz, and there he began the composition of
Euryanthe (J291). There were interruptions by visitors,
including Tieck, Jean Paul, Wilhelm Müller (who
dedicated the volume containing the *Schöne Müllerin*
and *Winterreise* poems to him) and Spontini, who came
with evidently friendly intent but whose notoriously
lofty manner upset Weber. His spirits, lowered by
illness, were not improved by the poor reception given
to *Preciosa* on 27 June; he finished his Fourth Piano
Sonata (J287), whose morbid background programme
was recorded by his pupil Benedict. In the autumn he
was again absorbed in theatre work (Morlacchi,

returned from five months holiday, was still inactive) and occasional music.

With the New Year of 1823, work on *Euryanthe* was intensified, and with it Weber became alarmed over the poor quality of Chezy's material. He refused to abandon her, however, and both demanded and contributed many revisions to the text. He conducted *Abu Hassan* on 10 March and *Fidelio* on 29 April, a performance in which Schröder laid the foundations of her later fame as Leonore. Other heavy demands kept him from resuming work on *Euryanthe*, and he requested an assistant in Gänsbacher; this proposal fell through, and eventually Marschner was appointed. There was another performance of *Der Freischütz*, with Schröder and her future brother-in-law Emil Devrient as Caspar, on 8 May, before Weber was able to leave for Hosterwitz, there to resume *Euryanthe*. Working for six or eight hours a day, he finished the sketches on 8 August and the whole opera, except for the overture, on the 29th.

On 16 September Weber left for Vienna with Julius Benedict. He arrived to find the city reacting against his music in the wake of several *Freischütz* parodies, and on the crest of a wave of Rossini enthusiasm (in which, against all his prejudices, he found himself sharing). He persuaded Henriette Sontag to accept the title role of *Euryanthe*, and settled the rest of the casting; he also found that he could count on support from various factions in the city. On 5 October he went out to Baden to visit Beethoven. The *Euryanthe* rehearsals, which had begun on the 3rd, went well: at first baffled by the work's novelties, the cast became enthusiastic and voluntarily extended their rehearsals. Schubert, who attended some of them, was not so impressed, though he did later commission *Rosamunde* from Chezy. The première, on 25 October, was reasonably well received; professional and critical reactions were mixed, and after Weber left for Dresden on 5 November audiences declined, despite the attempt by the new conductor, Conradin Kreutzer, to make the long work more popular by cutting it severely. After the 20th performance it was withdrawn.

When Weber reached Dresden again, by way of Prague, on 10 November, his disease had gained ground seriously: he was in continual pain, his limp was worse and he was coughing badly. Although he attended to his duties punctiliously, he was creatively at a standstill: between finishing *Euryanthe* and beginning *Oberon* over 15 months later he wrote only one little French song (J292). On 31 March 1824 he conducted *Euryanthe* in Dresden, with Schröder-Devrient; it was a success, but the effort made further inroads on his health, worsening his cough and obliging him to give up his daily walk.

Most of the early summer was spent quietly at Hosterwitz, though he did conduct Haydn's *The Seasons* and later Handel's *Messiah* for the Klopstock centenary at Quedlinburg: he was so overcome during 'I know that my Redeemer liveth' that he stopped conducting. A cure taken at Marienbad had little effect on his rapidly deteriorating health; back in Dresden, he set about cutting down his work and, haunted by presentiments of death, to gather together as much money as he could. On 18 August he received a letter from Charles Kemble, manager of Covent Garden, with an offer to write an opera for the 1825 season and to come to conduct both that and *Der Freischütz* in May, June and July. When he discovered from his doctor that he

23

could expect only a few years of life even if he spent them resting in Italy, he decided to accept the lucrative London offer for the sake of his family's future. Kemble then proposed *Faust* or *Oberon*; having forestalled Spohr with *Der Freischütz*, Weber now conceded his colleague *Faust*, and chose *Oberon*. Although the negotiations went slowly and intermittently, he began learning English, taking in all 153 lessons from an Englishman resident in Dresden, and achieved a fair command of the language.

Not until 30 December 1824 did he receive the first act of *Oberon* (J306) from the librettist chosen by Kemble, James Robinson Planché. Concerned at not being able to see the whole work at once, even in outline, he wrote to request postponement until the next season. On 6 January 1825 his second son was born, and christened Alexander. The remainder of the *Oberon* text arrived from Planché by 1 February, causing Weber considerable anxiety as he realized that this entertainment differed radically from his own idea of opera. But committed to the project, he accepted the text with the intention of later reworking it for the German stage. At this time he also set some Scottish folksongs (J295–304) for the Edinburgh publisher George Thomson. In the spring he took a new house at Cosels Garten, spending as much time as he could with his family (it was at this time that he sat for a portrait by Ferdinand Schimon). Yielding to medical pressure, he went in mid-July for a cure at Ems, where he was visited by Kemble and Sir George Smart, and was reassured about the business side of the *Oberon* project. He returned to Dresden at the end of August.

As a gesture to Spontini, Weber then staged *Olympie*

for a royal marriage, contributing some music of his own; and on 19 September, while preparing for this production, he resumed work on *Oberon*: Act 1 was finished on 18 November, Act 2 virtually finished by 3 December. On the 5th he went to Berlin to conduct *Euryanthe*, adding the pas de cinq on the 18th and directing the performance on the 23rd; to his sadness, he was unable to be home for Christmas. He left for Dresden again on the 29th, arriving to find his friends united behind Caroline in trying to deter him from going to London. His reply was that if he went, his family would be provided for, while if he stayed as an invalid, they would starve. He spent January 1826 on Act 3 of *Oberon*, and in making arrangements for the journey. Smart had offered him hospitality, and sent elaborate details to help him with the journey. Taking Anton Fürstenau as companion, he set off in a new carriage on 16 February.

V London and last illness

By a series of careful stages, which included a stop for two nights in Frankfurt, Weber and Fürstenau reached Paris on 25 February. Although he had planned to remain incognito, partly in order to avoid difficulties over the piracy of his works especially by Castil-Blaze, he paid calls on Auber, Paer, Catel, Cherubini and Rossini, among others. He attended various operas, including *Olympie*, and was so impressed by Boieldieu's *La dame blanche* that he immediately wrote to Theodor Hell recommending it. Berlioz, who spent a day seeking him, never found him; but he met Fétis, who thought his theories of tonality vague.

On 4 March he and Fürstenau reached London,

where he was comfortably lodged at Smart's house. On the 6th he visited Covent Garden, where he was warmly welcomed, and the next day he began rehearsals for his concerts. At the first of these, on the 8th, the entire audience rose to applaud him as he entered, and afterwards he was given a tumultuous reception. *Oberon* rehearsals began on the 9th: the cast was uneven, and included the famous tenor John Braham, a fine singer by then past his best but still a great draw. To suit him Weber reluctantly wrote 'O, 'tis a glorious sight to see' to replace 'From boyhood trained', also adding for him the *preghiera* 'Ruler of this awful hour' and 'I revel in hope and joy'. During the 16 *Oberon* rehearsals, though increasingly weak, he also gave as many concerts as he could so as to make as much money as possible out of the London visit. He did his best to meet the many demands, musical and social, which were put upon him, with a courtesy and courage that were widely admired. Although often barely able to get a glass to his mouth, let alone walk far without aid, he kept up his programme of engagements, in the letters home to Caroline, as to his public, pretending that he was less ill than the truth he admitted to his diary.

On 12 April, Weber conducted the première of *Oberon* (playbill shown in fig.3); although much of the applause was directed at the scenery and effects which played a large part in the entertainment devised by Planché, he was himself given the unprecedented honour of being called on to the stage at the end. In spite of his rapidly deteriorating condition, he continued to fulfil his engagements; Smart and his friends took him on excursions and lavished every care on him, but the unusually cold and foggy spring hastened his decline.

NEVER ACTED.

Theatre Royal, Covent-Garden,

This present WEDNESDAY, April 12, 1826,

Will be performed *(for the first time)* a Grand Romantic and Fairy OPERA, in three acts, (Founded on WIELAND's celebrated Poem) entitled

OBERON:
OR,
THE ELF-KING's OATH.

With entirely new Music, Scenery, Machinery, Dresses and Decorations.

The OVERTURE and the whole of the MUSIC composed by

CARL MARIA VON WEBER,

Who will preside this Evening in the Orchestra.

The CHORUS (under the direction of Mr. WATSON,) has been greatly augmented.
The DANCES composed by Mr. AUSTIN.
The Scenes painted by Mess. GRIEVE, PUGH, T. and W. GRIEVE, LUPPINO, and assistants.
The Machinery by Mr. E. SAUL. The Aerial Machinery, Transformations & Decorations by Mess. BRADWELL
The Dresses by Mr PALMER, Miss EGAN, and assistants.

Fairies.
Oberon, *King of the Fairies,* Mr. C. BLAND, Puck, Miss H. CAWSE,
Titania, *Queen of the Fairies,* Miss SMITH.

Franks.
Charlemagne, *King of the Franks,* Mr. AUSTIN,
Sir Huon, of Bourdeaux, *Duke of Guienne,*............Mr. BRAHAM,
Sherasmin, *his Squire,*................Mr. FAWCETT,

Arabians.
Haroun-Al-Rashchid, *Caliph of Bagdad,* Mr. CHAPMAN,
Baba-Khan, *a Saracenic Prince,* Mr. BAKER, Hassan, *Master of a Vessel,* Mr. J. ISAACS,
Ilamet, Mr. EVANS, Amrou, M. ATKINS,
Reiza, *Daughter of the Caliph,*...........Miss PATON,
Fatima, Madame VESTRIS,
Namouna, *Fatima's Grandmother,* Mrs. DAVENPORT.

Tunisians.
Almansor, *Emir of Tunis,*......Mr. COOPER,
Abdallah, *a Corsair,* Mr. HORREBOW, Slave, Mr. TINNEY,
Roshana, *Wife of Almansor,*...........Miss LACY,
Nadina, *a female Slave,* Mrs. WILSON.
Officers, Soldiers, Slaves, &c. of the different Courts.——Fairies, Sprites, &c.

Order of the Scenery :

OBERON'S BOWER,
With the VISION. Painted by Mr. Grieve.
Distant View of Bagdad, and the adjacent Country on the Banks of the Tigris,
By Sunset. Grieve
INTERIOR of NAMOUNA's COTTAGE. T. Grieve
VESTIBULE and TERRACE in the HAREM of the CALIPH, overlooking the Tigris. W. Grieve
GRAND BANQUETTING CHAMBER of HAROUN. F. Grieve
GARDENS of the PALACE. Pugh
PORT OF ASCALON. T. Grieve
RAVINE amongst the ROCKS of a DESOLATE ISLAND,
The Haunt of the Spirits of the Storm. Designed by Bradwell, and painted by Pugh.
Perforated Cavern on the Beach,
With the OCEAN—in a STORM—a CALM—by SUNSET—
Twilight—Starlight—and Moonlight. T. Grieve
Exterior of Gardener's House in the Pleasure Grounds of the Emir of Tunis. Grieve
Hall and Gallery in Almansor's Palace. W. Grieve
MYRTLE GROVE in the GARDENS of the EMIR. Pugh
GOLDEN SALOON in the KIOSK of ROSHANA. W. Grieve.
The Palace and Gardens, by Moonlight. Grieve.
COURT of the HAREM. Pugh.
HALL of ARMS in the Palace of Charlemagne. Grieve & Luppino

The Opera is published, & may be had in the Theatre, & of Mess. Hunt & Clarke, 38, Tavistock-street, Covent-garden

To which will be added (23d time) a NEW PIECE, in one act, called

THE SCAPE-GOAT.

Old Eustace, Mr. BLANCHARD, Charles, Mr. COOPER,
Ignatius Polyglot, Mr. W. FARREN, Robin, Mr. MEADOWS,
Molly Maggs, Miss JONES, Harriet, Miss A. JONES,

W. REYNOLDS, Printer, 9. Ironmonger-Court, Strand.

3. Playbill for the first performance of Weber's 'Oberon' at Covent Garden, 12 April 1826

27

He composed, partly by dictation to Fürstenau, a revision for wind band (J307) of his early march (J13) from the *Six petites pièces faciles* for piano duet, and finally (J308) the vocal line of a Moore setting, 'From Chindara's warbling fount I come': too weak to write down the piano part, he was nevertheless able to play it on 26 May.

By then he had become obsessed with thoughts of returning home and insisted on travelling directly (not, as originally proposed, through Paris) on 12 June. On 29 May he made his last appearance, at the benefit performance for the *Oberon* Reiza, Mary Anne Paton; afterwards, having reached Smart's house with difficulty, he advanced the departure date to 6 June. He refused to consider postponement, or to allow anyone to sleep with him in his room, and only with reluctance would he see a doctor. On the morning of the 5th, when the servants could not get an answer, his door was broken in and he was found dead. A doctor, summoned at about seven o'clock, estimated that he had been dead for five or six hours.

On 17 June a benefit concert was held for Weber's family, and on 21 June he was buried in Moorfields Chapel amid great public mourning. In 1844, Weber's successor at Dresden, Richard Wagner, arranged for the transfer of the coffin back to Germany. On 14 December it reached Dresden, and was conveyed to the cemetery to the accompaniment of Wagner's Funeral Music on themes from *Euryanthe*. The next day it was solemnly interred, with speeches by Theodor Hell and Wagner and the performance of Wagner's piece for male chorus, *An Webers Grabe*.

CHAPTER TWO

The man

I The critic

'The critic', wrote Weber, 'must with thoughtful discrimination and the utmost consideration support the progressive in art'. The cause nearest to his heart, and the one to which the centre of his life work was dedicated, was the evolution of German opera; and in all their variety, his writings uphold this and argue the case in different ways. He insisted that he was no mere chauvinist: 'I respect everything good, from whichever nation it comes'. But his aim was to encourage, by precept as well as practice, the kind of opera he declared Germans to want, 'a self-sufficient work of art in which every feature and every contribution by the related arts are moulded together in a certain way and dissolve, to form a new world'. He felt strongly enough about this definition of what Wagner was to call a *Gesamtkunstwerk* to repeat it from his review of Hoffmann's *Undine* (1817) in the fifth chapter (1819) of his unfinished novel *Tonkünstlers Leben*.

Weber's writings fall into several groups. Apart from some poetry (including ceremonial and occasional pieces, ingenuities such as acrostics and some tart epigrams), his principal imaginative work is *Tonkünstlers Leben*, on which he worked sporadically from 1809 to 1820. In some respects autobiographical, it is a kind of loosely constructed receptacle for ideas, satire, a romantic narrative, verses and other elements in a half-formed imaginative vision of an artist's life. Chapter 6 (1818) includes a scene at a masked ball in which national operatic genres are satirized, Italian, French and, sardonically and at some length, German too. Strongly influenced by Hoffmann, and to some extent by Tieck, the work anticipates Berlioz's more polished *Les soirées de l'orchestre*, as in the scene when the instruments of Weber's orchestra hold critical conversations among themselves (chapter 4). This section includes satirical comment on the latest Viennese symphony: there is no evidence that Weber was attacking Beethoven, as has been alleged (nor that he ever declared Beethoven 'ripe for the madhouse', a long-enduring fiction).

In early days 'somewhat precious and bombastic', as he admitted, Weber's style in *Tonkünstlers Leben* and in his mature criticism has a grace, wit and fluency that contrast agreeably with much critical writing of the day. He wrote a number of concert notices, in which, though he reviewed Hummel and Moscheles, he was principally concerned to encourage and draw attention to promising or little-known artists; but his most important critical writing is in a genre he developed as part of his systematic effort to widen understanding for German opera, the 'musico-dramatic notices written as an attempt, by means of historical descriptions and suggestions, to facilitate the appreciation of new operas'. He began these in Prague and continued them in Dresden, publishing them a few days before the performance. They include penetrating comment on operas by Spohr, Méhul, Boieldieu, Cherubini, Mozart (a sympathetic assessment of *Entführung*), Marschner, Meyerbeer (who is rebuked for abandoning the cause of German opera), and also reveal Weber's constant preoccupation with aesthetic and technical developments: in his notices of Dalayrac's *Macdonald* (the name by which *Léhéman* was known in Germany) (1811), Poissl's *Athalia* (1816) and Spohr's *Faust* (1816), he remarked on the valuable effect of musical motifs, which he described as 'the threads in the fabric of an opera'. In manner and in approach, these articles clearly anticipate Schumann.

Weber's more casual and isolated writings also reveal his steady preoccupation with the advancement of German music and musicians. While still a boy, he contributed entries to Gerber's *Neues historisch-biographisches Lexicon der Tonkünstler* when his own projection for a dictionary was abandoned. He formed and drew up the statutes for a so-called Harmonischer Verein that was to argue a case for the works of its members (who included Meyerbeer, Gänsbacher and Danzi). He planned a *Musikalische Topographie Deutschlands*, a kind of musical Baedeker for travelling musicians, and contributed an item on Basle; he also wrote reports on the state of music in various cities. His own works he seldom discussed, though there are informative essays on *Preciosa* and *Kampf und Sieg*; but some of his most valuable observations on word-setting are contained in a studied criticism he wrote of Friedrich Wieck's op.8 songs for the composer. His intensely practical side is further exemplified in his letter on the opera to Liebich in Prague, even more in his prospectus for the reorganization of the Dresden opera, in which he concerned himself with every detail down to the budget for the singers and their precise function in the ensemble. Throughout, his manner is elegant, knowledgeable and enthusiastic, generous and warmhearted (most openly in his letter of farewell to his pupil Julius Benedict), rarely revealing the pessimism that can be glimpsed in parts of *Tonkünstlers Leben*.

II The virtuoso pianist

In *Tonkünstlers Leben* Weber deplored the fact that 'these damned piano fingers, which through endless practising take on a kind of independence and mind of their own, are unconscious tyrants and despots of the art of creation'. Yet he was one of the great virtuosos of an age that included Hummel, Moscheles, Kalkbrenner and Czerny. Reviewing the first two artists, he selected as praiseworthy qualities musical honesty, clarity and a singing tone, making little mention of technical brilliance. Certainly Weber himself possessed abundant brilliance, as is evidenced by his shorter salon pieces, by his habit of loosening up his fingers by playing the difficult C major finale of his First Piano Sonata at full speed in C♯ and by the accounts of his contemporaries. His Berlin friend Hinrich Lichtenstein also described how he had the gift of making everyone feel that they were hearing a work for the first time, and how, while Hummel and Kalkbrenner's improvisations seemed out to please, 'with Weber the impression at such times was that he had above all found a means of revealing his deepest feelings to his closest friends, and that his whole being was concentrated on making himself understood'. Lichtenstein further testified that 'Weber was as great a master of the guitar as of the piano'; with this instrument, as a review of a concert by Giuliani testifies, he also admired a singing tone.

One of the great orchestrators of the day, Weber was also aware of the new tonal possibilities of the piano, especially of the Viennese instrument he favoured (a Brodmann). This was lighter than a modern instrument, and its keys were narrower, with an octave span of 15·9 cm as against the modern 16·5 cm. Moreover, an engraving shows that Weber had long fingers and an exceptionally elongated thumb that reached to the middle joint of his index finger: this enabled him to play four-part chords covering a 10th without difficulty, and to devise for himself spectacular leaps and the octave glissando of the *Konzertstück*. As Benedict described it, 'Having the advantage of a very large hand, and being able to play 10ths with the same facility as octaves, Weber produced the most startling effects of sonority and possessed the power, like Rubinstein, to elicit an almost vocal quality of tone where delicacy or deep expression were required'. Much of his piano music reveals not only a relish of piano tone – as with the opening flourishes of the Second Piano Concerto – but the love of a clear, singing melody supported by lightly strummed chords. Lichtenstein's account indicates that his wish was above all to communicate feeling, rather than merely to impress, as part of his process of giving a new Romantic role to the virtuoso – as a man among equals, speaking, by his greater sensibility and greater gifts, of emotions common to all.

III The Kapellmeister

Weber's virtuosity extended to his conducting. He belonged to the first generation who directed the orchestra standing and with a baton or a roll of paper, normally grasped by its middle. The Hayter engraving (see fig.4) shows this, and was evidently a vivid representation of his manner: 'It conveys to the spectator

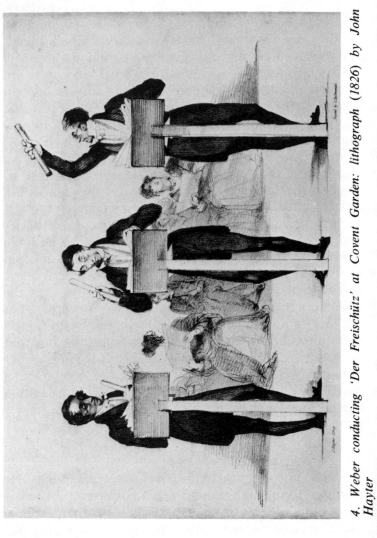

4. *Weber conducting 'Der Freischüz' at Covent Garden: lithograph (1826) by John Hayter*

a perfect general impression of his countenance, look, and appearance. The central figure is the most striking resemblance; though that on the left is also very good. . . . It makes Weber too tall . . . but it is like, and affords a perfect notion of the original, living as he led' (*Literary Gazette*, London, 17 June 1826, p.380). The animation and the infectious enthusiasm of his conducting are elsewhere praised, as well as his quick control of the orchestra. Conducting Haydn's *The Seasons* in 1824, he was gratified to find that 'I had the marvellous feeling of being able to express myself with my orchestra as completely as if I were sitting at my piano alone and playing what I liked'; and Wilhelm Müller noted in his diary of this performance, 'In conducting Weber became so moved and excited that he stood as if transfigured and often seemed to be making music with his whole face'. He disliked rigidity of tempo: though opposed to the metronome in principle, regarding it as an attempted substitute for musical feeling, he did suggest metronome markings for *Euryanthe* but at the same time warned that every slow passage may include faster movement and vice versa.

In his wish to obtain better blend and control, he reorganized the conventional seating of the opera orchestra. At Carlsruhe, his first appointment, he redistributed the players so as to achieve greater homogeneity of tone; and in Prague he showed his appreciation of the value of a good leader in engaging Franz Clement from Vienna. On his arrival in Dresden he found the orchestra seated in the haphazard arrangement that had endured from Hasse's days: the conductor was at the piano in the middle, unable to see some of the players and with a cello and bass reading the score over his shoulders. As Weber pointed out in his letter (21 January 1818) to Count Vitzthum, this was a legacy of continuo playing that was hardly relevant to the needs of Spontini or Cherubini or the other new works he intended to introduce. His plan involved moving the upper strings to his right, the wind and brass to his left, allowing the lower strings to remain behind him as he wished to move his desk nearer the prompter's box.

Such an arrangement, leaving some of the orchestra out of his sight, might seem little improvement; but as Weber pointed out in the same letter, the opera director's place was in the centre of the performance, controlling both orchestra and stage, and the new position allowed him to keep in close touch with his singers, to see into the wings and even to have unobtrusive contact with the prompter. Such a total control lay at the centre of his whole approach to opera. From the moment of his arrival in Dresden he tried to put into operation methods for which he had been preparing since his earliest days. Already at Carlsruhe he had annoyed the singers by insisting on group rehearsals and by giving what they considered exaggerated attention to the orchestra, as well as by introducing new works that fitted his ideas for the future of opera. And in Prague, he had not only coached the singers himself but had supervised every detail of production, scenery, dances and costumes. His position at Dresden, and the support of Vitzthum, encouraged him to go even further in his pursuit of the unified work of art. He cast or recast the company carefully, with an ear to ensemble rather than to vocal virtuosity. The soloists he impressed with the need to understand the dramatic and moral import of their roles; he was delighted with the arrival of

Wilhelmine Schröder and the revelation of her dramatic talent that was to arouse Wagner to a sense of vocation. In his own operas he emphasized that Caspar should be acted rather than sung, and that 'She who cannot sing the last passages of Eglantine's aria with blazing fire had better simplify the number so that the passion of the whole is not chilled' (in the article on the *Euryanthe* metronome markings). He also placed new stress on the role of the chorus, which previously had made little more than a token contribution to the drama.

No less importance was attached to the visual side of the productions. He asked for a subtler approach to costume than the dressing of villains in black and heroines in white, and suggested that (as in the case of Lysiart) the contrast between a noble exterior and inner corruption might be dramatically potent. He reformed the lighting, importing some new Argand lamps to replace the old candles with the brighter and more flexible device of cylindrical wick and glass. The torn and sooty scenery which these lamps illuminated was replaced, and the décor was now conceived as mobile painting, playing a functional expressive role in the drama. His strong visual sense is also evident in his constant use of painting metaphors: he asked singers to 'portray' their roles with 'fine brush-strokes' or more broadly 'like a fresco'. He turned the Dresden fondness for set tableaux to new expressive effect by making these a structural feature of his productions, conceiving them as part of the design to halt and intensify the drama at certain points rather than as mere decoration.

In order to impress his ideas more effectively upon singers little expecting such a radical approach to what they saw as a purely vocal art, he instituted a novel system of rehearsal. He would begin with a *Leseprobe*, at which he would read and explain the text (so vividly that he was lightheartedly offered a job by a Vienna stage manager). In 1825 he anticipated the modern Dramaturg when he appointed Tieck to be a so-called *Literator*, helping singers with meaning, pronunciation of foreign words and other details, as well as reading new texts. After the *Leseprobe* would come a *Zimmerprobe* with each singer (now the work of a répétiteur), then two *Setzproben* at which the mise-en-scène was translated into action, and lastly two to four *Generalproben*.

The repertory which Weber, thus carefully prepared, introduced in Prague and Dresden was chosen to set the stage for the appearance of German Romantic opera. He saw little historical foundation for significant creative German opera, which had needed Beethoven or Mozart to make universal masterpieces of Singspiel. He resisted what he regarded as the shallowness of Italian opera, with its concentration on vocal superficies. It was chiefly in French opera that he discerned the foundation of Romantic opera; and his repertories show a practical understanding of what the French had to offer German audiences and German musicians. In Prague he opened with Spontini's *Fernand Cortez*, following this with a dozen works by Spontini, Cherubini, Isouard, Boieldieu and Dalayrac. The only German work in the first six months was Fränzl's popular *Carlo Fioras*; the only Italian operas which he ever produced there were a German version of Fioravanti's *Le cantatrici villane* and works by Salieri and Paer (both of whom worked in Germany or Austria). Other operas to appear in the Prague repertory were by Grétry, Méhul and Catel;

Weber never produced one of his own works there. In Dresden the repertory was similar, and included works by Cherubini and Méhul, as well as a larger proportion of German works, among them *Fidelio*, five works by Mozart, Spohr's *Jessonda*, Weigl's *Die Schweizerfamilie*, Winter's *Das unterbrochene Opferfest*, and later his own *Freischütz* and *Euryanthe*. The rivalry of Morlacchi's court-supported Italian opera and the generally uneducated Dresden taste, which leant towards vaude-villes, farces and melodramas for entertainment in the vernacular, make this achievement seem even more impressive. Dresden was one of the most conservative of all German cities, and yet it was there that the most searching and imaginative steps forward in German opera were taken.

CHAPTER THREE

Works

I Operas and other stage music

'I am waiting in agony for a good libretto', Weber wrote to Gänsbacher in 1811. 'I don't feel right when I haven't got an opera in hand.' He once even advertised for a libretto, but despite his lifelong interest in opera he never found a librettist worthy of his talents. In early 19th-century Germany, libretto writing was considered, despite Goethe's Singspiels, a craft beneath the attention of the major writer, best handled by men with practical theatrical experience. Poissl (whom Weber admired) was almost the only German composer before Lortzing to make a regular practice of writing his own librettos. But Weber was highly critical, and never shy of taking unprecedented steps; he also possessed one of the liveliest pens in Germany, and thus it is puzzling that he never seems to have considered writing for himself. It seems most probable that his sense of theatrical effect allowed him to suspend his better literary judgment and to entrust himself to inferior writers because of their stage experience; but his attempt to break out in a new direction with *Euryanthe*, for which he wanted an unorthodox libretto and deliberately chose a poet with no theatrical experience, was wrecked on the feeble talent of Helmina von Chezy.

With *Euryanthe* Weber was moving in the direction of continuous opera demanding texts of a kind without real German precedent. He intended to rewrite *Oberon* as an opera with recitatives for the German stage when he returned from England; and there is a subtle and pervasive use of motivic ideas, even in the disjointed numbers imposed upon him by English musical stage conventions, which shows that he was ready for a much larger imaginative enterprise. The term 'Leitmotiv' was coined by F. W. Jähns to cover the practice evolved by Weber; and although Jähns applied it loosely, he was right to perceive the beginnings and growth of a system for which the old descriptions *Reminiszenzmotiv* or *Erinnerungsmotiv* were inadequate. The Wolf's Glen scene (*Freischütz*) freely uses motifs from earlier scenes and is harmonically organized by a key sequence based on the diminished 7th chord F♯–A–C–E♭ associated with Samiel and the powers of evil. *Euryanthe* contains the most advanced example of true leitmotif then to have appeared in opera. Associated with the 'anti-heroine' Eglantine, it anticipates Wagnerian practice and is associated not simply with Eglantine but with her deceit: when she takes a part openly against Euryanthe, it is not further used. Previously it has been modified from its original appearance to reflect various processes of her attempt to win Euryanthe's secret, wheedling, cajoling, pretending affronted innocence and so forth. It is used not only in direct allusion, but casually and as the material for the accompaniment of Eglantine's vocal line (see ex.1, p.42).

Euryanthe projects across a whole opera the mastery of composing formally organized sections, recitative, freely discursive passages and motivic references, which Weber had achieved in the Wolf's Glen scene. Earlier examples of continuously composed operas are not hard

41

Ex.1

Moderato assai

lusingando dolcissimo

altered as

tranquillo

p   etc

allegro

pp   etc

to find: *Euryanthe* enlarges the range, extending it beyond anything previously achieved in German opera
and looking forward to and beyond *Lohengrin*. Weber's
developing concern for unity and continuity is, paradox-
ically, also shown in *Oberon*; a sense of Oberon's
musical presence is stamped on musical numbers
separated by long stretches of dialogue by the use of the
rising three-note figure of the overture's opening, quoted
from oriental dances discovered by Weber (see ex.2), to
suffuse the invention of the entire opera. Some of the
many ways in which this figure occurs, sometimes
inverted, are shown in ex.3.

Although in varying degrees they retain traces of
their Singspiel origins, Weber's earlier operas show a

steady progress towards the larger ideal of German Romantic opera of his maturity. Nothing is known of his first attempt, *Die Macht der Liebe und des Weins*, written when he was 11; and of *Das Waldmädchen*, only two fragments survive apart from whatever was re-used in *Silvana*. Weber himself remembered *Das Waldmädchen* as, 'a very immature piece, though perhaps here and there not a completely uninventive one' (*Autobiographische Skizze*). There are few touches of individuality in the fragments, but the composition gave Weber a glimpse of Romantic opera, one which broadened considerably the next year in *Peter Schmoll*.

Ex.2 Oriental dances discovered by Weber

Ex.3

Based on a contemporary novel of little merit, the *Peter Schmoll* libretto retains from its source certain features, common to the lesser writing of the time, which were to survive into *Der Freischütz* – a feeling for nature, a reliance on fateful coincidence, the intervention of a holy man to resolve the situation, and a tendency for the lesser, bourgeois characters to assume fuller prominence and life than the ostensible leading characters. For all its occasional bluntness and its limited range, *Peter Schmoll* is effective and remarkably forward-looking for an inexperienced 15-year-old composer. The use of expressive key patterns anticipates *Der Freischütz*, as does the portrayal of the heroine (an Aennchen with the introspective bent of an Agathe). She is the only female role; the men, apart from the weakly characterized hero, are drawn with a quick sense of dramatic variety and an instinctive flair for melody derived from the contours of German folk music. There is also an acute, fresh use of instruments, including such rarities as recorders and basset-horns.

Weber described *Silvana*, like its progenitor *Das Waldmädchen*, as a 'Romantic opera'; and despite the unevenness that partly derives from the re-use of earlier music, it moves deeper into the Romantic territory he was to make his own with *Der Freischütz* and *Euryanthe*. The world and conventions of medieval chivalry and the charm of the open-air life are both invoked in a story dealing with the wooing and winning by a knight, against many difficulties, of a girl he discovers living in the woods. Her mysterious charm is increased (and the plot made easier) by her muteness; but this set problems to the composer that were only partly solved by his rapidly developing sense of the expressive potential of the orchestra. Rudolf, the hero, is no less vivid than Huon in *Oberon*, whom he foreshadows; while the villain Adelhart is an ambitious piece of character-drawing, a vehement bass whose villainy, as with Lysiart, does not exclude a touch of humanity. Although other features of the work are conventional, the direction in which Weber's interests were tending shows in the skilful handling of the orchestra to depict the forest, in sunlight or storm, and its inhabitants (including a chorus of huntsmen). A revision made by Weber in 1812 for Berlin tightened up some of the arias and made other improvements. He was by then aware of a lack of variety in the work; he already had behind him the 'much clearer and sounder' *Abu Hassan*.

By description and by intent, *Abu Hassan* is a Singspiel, a one-act frolic that surpasses the routine humours of most contemporary examples by the wit and human tenderness in its music. The whole piece reflects the security Weber had gained as a composer in the wake of his Stuttgart period and his self-discovery through his new friends in Mannheim and Darmstadt. The scoring is no less resourceful than before, but geared to a quick-witted musical invention that is at its best worthy of comparison with Mozart's *Entführung*. Weber's lifelong tendency to conceive his melodies in instrumental terms here finds expression in lively, neat contrasts between his hero and heroine, Abu Hassan and Fatima, as they veer between banter and affection. The aria Weber later added for Fatima, 'Hier liegt', a mock lament composed at the time of *Euryanthe*, is a fine piece with a useful dramatic function, but it bears rather heavily on what is essentially the lightest, most deftly sketched of operas. This does not exclude some

45

5. Autograph MS of
the beginning of the
finale of Weber's
'Der Freischütz',
composed 1817–21

subtle musical strokes, as when the chorus of creditors mollify their music from tough counterpoint into placid block harmony to support Abu Hassan's exuberant melodic flourishes of relief at having talked them round. The charm and wit of *Abu Hassan* make it particularly regrettable that Weber never completed another comedy.

In its outward form *Der Freischütz* is a Singspiel based for the most part on separate numbers, many of them close in character to popular song, linked by speech. However, Weber combined a number of factors to give the work a new creative status in the tradition. The fluent structure of the Wolf's Glen not only draws skilfully on motifs, a sure dramatic instinct for the balance of freely and tightly organized musical sections and a carefully planned overall harmonic design, but also introduces linking dialogue and, with particularly telling effect, speech used as melodrama. He pointed out to J. C. Lobe that 'half the opera plays in darkness. In the first act it is evening and its second half plays in the dark; in the second, it is night during Agathe's big scena, with moonlight through the window, and finally at midnight come the apparitions in the Wolf's Glen. These dark forms of the outer world are underlined and strengthened in the musical forms'. This movement from light into darkness and out into light again, reflecting the advance and retreat of the powers of evil, is reflected not only in the visual designs and lighting Weber intended for the opera, but also in the harmonic plan he devised for the music. The association of the quintessentially Romantic chord of the diminished 7th (a chord without a root and harmonically equivocal) with Samiel draws the music of the Wolf's Glen scene into an area of threatening uncertainty, sharply antithetical to the robust key structure associating C major with the powers of good and D major with the healthy natural world in which the events take place. Reinforcing this is a careful pattern of instrumentation. As Weber also told Lobe, he devised tone-colours to match the two principal elements he identified as 'hunting life and the rule of demonic powers'; the horns were the obvious choice for the former, for the latter, 'the lowest register of the violins, violas and basses, particularly the lowest register of the clarinet, . . . bassoon, the lowest notes of the horns, the hollow rolls of drums or single hollow strokes'. The actual tunes for the village and forest life Weber admitted he had based on his study of folk music. In some cases, actual folktunes figure behind or within his melodies, but so fully absorbed as to take on his own individuality, and even in turn to find their way into currency as German popular songs still seen in student songbooks.

The gift of creatively drawing on earlier example, which Weber shared with all other important composers, also shows in his characters. He looked to French opera, not Singspiel, as the model for German Romantic opera; Max has his ancestry principally in Méhul, Caspar (by way of Pizarro and Weber's own Adelhart) in Cherubini's Dourlinski, while Aennchen is closer to the soubrette of *opéra comique* than to the earthier Singspiel servant types. Yet Weber gave them, and all his characters, a vivid, peculiarly German colour, so that in 1821 an entire nation was able to recognize an essential part of its nature in these figures and this setting, caught in music of memorable appeal.

Naturally Weber was urged to repeat so striking a

success, but he was anxious to move beyond a work whose wild popularity soon came to irk him. Although keen to retain in *Euryanthe* a supernatural element for which he knew he had a special touch, he saw in the opposed pairs of Adolar and Euryanthe, Lysiart and Eglantine, a novel kind of dramatic contrast based on moral grounds; both the advance in his musical language and his ambitions for German opera demanded something more elaborate and large-scale than *Der Freischütz*. His harmonic language had developed rapidly, and the chromatic freedom of certain passages in *Euryanthe* was not overtaken until Liszt and Wagner. The opera's harmonic structure is based more on the antithesis between diatonic and chromatic harmony than on key references, though these still play a part, with E♭ for the courtly tonality and sharper keys for the villains. This allowed Weber to move a long way towards continuous, freely composed opera. The reliance on closed forms is not wholly shed, but the formal arias are contained within larger structures or linked, with varying degrees of fluency, to recitative. The opera's melodic style similarly ranges from formal constructions to conventional recitative, but is most remarkable in passages of free yet well-organized declamation, anticipating Wagner's endless melody. Eglantine's highly flexible motif provides the basis for the most striking of these passages.

The weakening of the set aria in *Euryanthe* further transfers the burden of expression from the singer to the orchestra, which is used with even greater richness and subtlety than in *Der Freischütz*. There is a new expressiveness in the string writing, especially in the characteristic use of individual colours as with the eight muted solo violins of the overture against an unmuted viola tremolo figure, the viola solo against clarinets and bassoons in no.12, or the two solo cellos over unison cellos and basses in no.5. But these instances are not merely effects, for Weber was as always concerned to unify his operatic resources: they arise from a precise dramatic need, as does the harsh, thick doubling in the wedding march for the evil pair, or the unaccompanied bassoon opening no.17, characterizing in its curious melody and its high, plaintive tone Euryanthe's isolation and despair.

For all the weaknesses in the libretto, which have been heavily emphasized over the years, *Euryanthe* is a mature and advanced music drama, one which was deeply admired by later Romantic composers including Schumann, Liszt and Wagner, and which placed Weber in a commanding position for greater achievements. His mortal illness, and the need to work in the stultifying context of English pantomime opera forced on him by Covent Garden and Planché, compelled him to refine and distil his idiom into separate numbers for *Oberon*. The amount he achieved is, in the circumstances, astonishing: apart from its subtle motivic handling, *Oberon* uses, especially in the woodwind, delicate textures that were to be a model for Mendelssohn and Berlioz in their fairy music. His response to the pervasive element of the sea (which he had only once experienced in his life) was hardly less poetic than his earlier depiction of the forest, whether in the gentle lulling of the ravishing Mermaids' Song, or in the storm, or in the famous apostrophe of Reiza, 'Ocean! thou mighty monster'. Even in this disjointed and inevitably uneven work, Weber's aural preciseness fixed on inventions that were to ring in the ears of later composers, not least Wagner from *Der fliegende Holländer* (the Dutchman's first aria echoing Oberon's) to *Siegfried* (Brünnhilde's awakening echoing Reiza's hailing of the sun). As with *Der Freischütz* and *Euryanthe*, the overture to *Oberon* concentrates some of the most important elements of the musical drama into a loose sonata structure. Weber's expressive intent in his opera overtures was not to provide a potpourri of the best tunes, or merely a separate piece to settle the audience, or even a summary of the drama, but to use music from the opera in a work of art that is self-contained and at the same time presents the principal dramatic conflicts as they are characterized musically.

The comic opera *Die drei Pintos* was finally abandoned in November 1821 with only seven numbers sketched (totalling 1769 bars). They include some promising ideas for a lively comedy, which led Caroline to approach Meyerbeer. When he returned them 20 years later, having been unable to do anything with them, they passed into the hands of Mahler, who knew the Weber family in his Leipzig years. By drawing on a wide variety of little-known Weber pieces, he worked them up into a viable comic opera. Only the entr'acte and no.20 (Finale A) are fully composed by Mahler from Weber material; the rest consists, to varying degrees, of added accompaniments, filled-in links, some development and the orchestration of both sketches and imported music.

Of Weber's other stage music, the most substantial item is the incidental music for P. A. Wolff's drama *Preciosa*. Although he took the trouble to study Spanish and even gypsy music, Weber had no more direct con-

tact with Spain than hearing some soldiers' songs with
Spohr, and the flavour of the *Preciosa* music is far more
German than Spanish. However, it makes skilful use
of melodrama, in a manner anticipating the Wolf's Glen,
and the melodic charm of the songs and choruses won
Weber some useful popularity in Berlin a few months
before the première of *Der Freischütz*. His incidental
music for Schiller's *Turandot* consists of an overture
and six numbers, all based on an *air chinois* he found in
Rousseau's *Dictionnaire de musique*. The most remarkable number, after the overture, is the interestingly
scored funeral march. The numbers he wrote in
profusion for insertion in other composers' operas and
in plays are mostly short, often just a few bars of
characteristic music or a single song; some found their
way into more famous works, as with that for Eduard
Gehe's *Heinrich IV*, largely re-used in *Preciosa* and
Oberon.

II Choral music
Weber's sacred choral music consists of three settings
of the Mass (the second and third with an offertory
added at a later date) and a setting of the Agnus
Dei for use in a play. Lost from the time of its composition in 1802 until it reappeared in Salzburg in 1925, the
so-called *Grosse Jugendmesse* was evidently abandoned
by Weber as several passages recur in the later E♭ Mass.
It is symmetrically constructed around a central Credo,
a kind of operatic scena for bass, with the first and last
numbers using identical music, and with a similar
distribution of numbers lying between each of them and
the Credo. The music does not live up to the originality
of the scheme. However, the E♭ *Missa sancta* of 1818,

though uneven partly because of its use of earlier music including three of the six fughettas (J1), is a work of considerable ingenuity and drama. The Sanctus opens with some isolated choral and orchestral chords clearly intended to make use of the Dresden church's reverberant acoustic in a manner Berlioz was later to adopt; and at the king's request, there are passages, notably a beautiful Benedictus, designed to display the voice of the much-admired resident castrato, Giovanni Sassaroli. The richest writing is in the final Agnus Dei, a movement that vividly suggests the darkness out of which the prayer is raised. The Mass in G is less dramatically cast, and being written for the royal golden wedding is celebratory and openly tuneful. It is also more unified and even than its predecessor. The Credo is organized as a series of bold choral entries against the working out of a motif on the violins. The operatic quality which Weber found hard to dismiss from his sacred music shows pleasantly in the Benedictus for four soloists and in the charming Agnus Dei for contralto. In each mass an offertory designed to show off Sassaroli's virtuosity fulfils its purpose well enough.

Weber's secular choral music can be categorized, though not formally separated, into cantatas, accompanied choral works, and unaccompanied choral works and partsongs. With a composer for whom form and medium were often empirical matters, the borderlines are indistinct. Thus, the cantatas range from a comparatively small and uninteresting piece such as *In seiner Ordnung schafft der Herr* to large-scale cantatas, and include the hybrid *Der erste Ton* for reciter and orchestra with chorus, which retains a certain interest for its early use of melodrama, its use of the diminished

7th as a unifying feature of motivic force, and some remarkable anticipations of Wagner. *L'accoglienza* is an occasional work, drawing on previous music and in turn being quarried for later works. The *Jubel-Kantate* is another occasional, celebratory piece, efficient but showing signs of haste in its lack of distinctive ideas. By far the largest cantata is *Kampf und Sieg*, written in the wake of Waterloo. As Weber pointed out in a long essay analysing the work reprinted as no.75 in Kaiser, his intention in omitting arias was to keep the drama of battle and victory moving before the listener's ear without interruption; there are accordingly only two extended numbers, both admirable instrumental tone pictures. Elsewhere, in shorter choral and orchestral numbers, the emotions of battle are described, with prayers before battle, authentic marches and Prussian horn calls, and other ingredients; they culminate in a routing of the French *Ça ira* with music that includes *God Save the King*, leaving the field to pious sentiments on the merits of honourable victory and peace. For all its crudity it is an effective work, a kind of concert opera foreshadowing some of the techniques Berlioz was to use in *La damnation de Faust* and suggesting how much was lost in the years during which Weber had no time to write an opera.

Weber's enthusiasm for the patriotic spirit of the times also shows in the cycle *Leyer und Schwert*. Although these volumes contain chiefly solo songs, they include six male-voice choruses (vol.ii), among them the once enormously popular *Lützows wilde Jagd*. Like most of Weber's choruses, they derive closely from the 'gesellige Lieder' which have always been part of the German beer-cellar repertory. They contain pious numbers that verge on the sentimental, of which a distinguished example is the *Gebet vor der Schlacht*, but there are tuneful, vigorous, even hearty numbers that are as suited to a student drinking evening as to the *Liedertafel* in Berlin at which Weber presided. Their style also provided the creative background for the Huntsmen's Chorus and similar operatic items. Weber's idiom, in planting its roots in German popular music, drew from town as well as country, from the beer-cellar and amateur choral society as well as the peasants' songs.

III Songs

Weber's songs include the majority of the patriotic *Leyer und Schwert* collection, though normally he did not write solo songs in order to stir up popular emotion. In many cases they reflect his private life or activity with friends. The Darmstadt and Heidelberg days encouraged some charming guitar songs; his Swiss trip and contacts with singers led to a few Italian canzonettas; and near the end of his life he contributed ten arrangements of Scottish folksongs to George Thomson's collection. But most of his large output of songs consists of German lieder for voice and piano. His flowering as a songwriter coincided with the discoveries of Arnim and Brentano in *Des Knaben Wunderhorn*, and he was naturally drawn to that collection's revelation of the virtues of folk melody. He was not, however, interested in exploring the richly detailed, atmospheric scene-setting in the piano and its reflection and support of the poetic mood, which became the especial glory of German lied. Though capable in certain songs of a lively thumbnail sketch, as with the bagpipe 5ths, the wrong notes and the clumping clogs of the wedding

6. Carl Maria von Weber: portrait by Ferdinand Schimon (1797–1857)

dance in *Reigen*, he was mostly concerned with evoking a general mood and with creating an effective and well-articulated vocal line. In some valuable comments made to Friedrich Wieck, who had sent him some songs for criticism, Weber declared that 'the creation of a new form must arise from the poem one is composing. In my songs I have always made the greatest efforts to render my poet truly and correctly declaimed so as to produce new melodic forms'. It was on accuracy of declamation rather than on depth of meaning which he fastened; and he never looked to great poetry for his texts, his literary sense leading him to regard it as self-sufficient rather than as material for music.

Many of his songs are in varied strophic form. They reveal a gift for melodies that may look plain on the page until sympathetic performance reveals their accurately judged grace or charm. He understood how strophic variation can develop a fruitful dramatic and expressive tension between the expected and actual repetition; for him this was almost invariably an expression of the poem's emotional development rather than a description of the scene.

The finest of these comparatively straightforward yet delightful songs suggest that Weber might have become the 'master of German song', as Wilhelm Müller hailed him. *Was zieht zu deinem Zauberkreise* (J68) is almost Schubertian in its passionate involvement with the poem and its beautifully judged interplay between voice and piano. *Die Temperamente beim Verluste der Geliebten* (J200–03) are a group of four songs that virtually constitute a cycle, along with Beethoven's *An die ferne Geliebte* the first in the history of the lied, and their exploration of four different emotions at the loss of a lover is done with wit, passion and elegance. Weber ranged wide in his methods, occasions and media for song, often using it to catch in music a particular emotion of a particular time or place. It is the very diversity of his activity in this sphere that has contributed to his being underrated.

IV   Orchestral music

Weber's orchestral output includes three concert overtures. The *Grande ouverture à plusieurs instruments*
(J54) is a reworking of the overture to *Peter Schmoll*.
Der Beherrscher der Geister (J122) was described by
Weber as 'a completely new working' of the lost
Rübezahl overture, and is one of his most brilliant
pieces, alternating a hurtling main theme treated with
contrapuntal virtuosity and a darker central section for
brass choir and drums, satirically likened by Weber to
'an artillery park'; however, he was justly pleased with
the work's 'strength and clarity' (letter of 15 November
1811). The *Jubel-Ouverture* (J245) has no musical
connection with the contemporaneous *Jubel-Kantate*,
and consists of a celebratory Adagio leading to a lively
Presto and culminating in a full-scale orchestration of
'Heil dir im Siegerkranz' (*God Save the King*), imported
from *Kampf und Sieg*.

With two unimportant exceptions, Weber's only
other purely orchestral works are his two symphonies.
Written during his Carlsruhe days for the ducal
orchestra, they show little feeling for the genre, and
indeed Weber himself was the first to recognize their
formal deficiencies, suggesting that the opening
movement of the First Symphony was more like an
overture than a true symphonic movement. However, in
the vigorous thrust of the finales and especially in the
dark-hued scoring of the slow movements, they reflect
the imagination that Weber was soon to channel into
more sympathetic enterprises.

The lack of conviction in the symphonies is significant. In most of his concertante works Weber wrote the
slow movement and the finale before tackling the first

58

movement; and with the *Andante e Rondo ungarese* for viola (J79, revised for bassoon as J158) and the *Adagio und Rondo* (J115) for harmonichord, he in effect simply left the first movement out. However skilfully he matched himself to the demands of sonata form in his concertos and sonatas, the true quality of his invention always shows in the grave, imaginatively constructed and scored slow movements and in the brilliant finales, whereas the first movements are to be measured more by the degree to which they can accommodate such qualities than by the mastery of form that comes from real belief in it. He was most successful with instrumental music either when drawing on a dramatic idea, or when working in an empirical form also perhaps derived from a programme. He was reluctant to disclose such programmes, fearing reasonably that the works would then be regarded as purely illustrative. The well-known 'plot' of the *Konzertstück* for piano, revealed by Benedict, about a returning crusader was the prompt for a new form rather than the subject of a tone poem. As such, it encouraged later Romantic composers to base their concertante works on dramatic ideas, and to develop new forms which would replace the sonata's domination and still be satisfying in strictly musical terms.

The Horn Concertino is basically an introduction and a set of variations, into which is inserted a cadenza in the form of a recitative that includes a passage making use of the horn player's trick of sounding chords; the finale is in the popular Romantic polacca rhythm that Weber often used. The Clarinet Concertino, the first of his works for Heinrich Baermann, similarly has a slow introduction leading to a set of variations with a passage before the final Allegro ingeniously matching the clarinet's lowest register to one of Weber's favourite orchestral sounds, violas in 3rds. Even in the three full-scale wind concertos, despite the deftly composed first movements, it is in slow movement and finale that his most characteristic voice is heard. With his dramatic sense of timbre, he was able to 'personalize' the instruments and to reveal new aspects of their tonal characteristics: in the First Clarinet Concerto and Bassoon Concerto, he included in each of the slow movements a passage in which the soloist intones against solemn horn chords, and in the Second Clarinet Concerto and Bassoon Concerto he gave the soloist an openly operatic character with prominent recitative passages. With the finales, he was concerned above all to demonstrate brilliance, no less so in the witty Bassoon Concerto than in the clarinet works.

Weber's piano concertos, written for his own virtuosity, are similarly rich in passages which delight in novel instrumental combinations. Each slow movement opens with the piano set against unusually scored strings, in no.1 violas, two solo cellos and basses, in no.2 muted four-part violins and solo unmuted viola (anticipating *Euryanthe*); and each movement includes some pearly runs and elaborate, almost improvisatory chromatic flourishes that take the music well into Chopinesque territory. The first movements include different aspects of his love of piano tone, no.1 introducing an almost vocal melody over lightly strummed chords, clearly designed to show off a singing tone, no.2 opening on tremendous arpeggio flourishes, while the finales give plenty of opportunity for display to the composer–virtuoso's finger dexterity. There are passages in the concertos which suggest in their spectacular leaps that he was expected to be seen as well as heard overcoming fearful difficulties. Certain sections, and some whole movements such as the fine Adagio of the Second Piano Concerto, are virtually colouristic studies, and show how much inventive music Weber was able to extract by means of his individual approach to the problems of instrumental concerto writing; many of his original ideas of piano technique and sonority were taken up and developed by later pianist–composers, especially Schumann, Mendelssohn and Chopin. For Liszt the inspiration was at least as much in the novelty of form suggested by the *Konzertstück*.

V Chamber music

As a composer whose instincts were primarily dramatic, Weber did not turn naturally to the essentially private art of chamber music (he never seems to have contemplated a string quartet); all his chamber works include his own instrument with the sole exception of the Clarinet Quintet which was written for Baermann, an engaging but slender work that is really something of a chamber concerto. Much the same is true of his Piano Quartet, whose most striking movements are, characteristically, the Adagio and especially the inventive, brilliant finale. The Trio for flute, cello and piano is a much larger creative enterprise: its slow movement is probably a piece written in Prague and recast as an eloquent and powerful Andante whose contrasts of anxiety and even despair, set against a more resigned melancholy, colour the movements added later. It is exceptional in his output in trying to contain extremes of Romantic emotion within a fairly strict Classical framework; the opening Allegro achieves this balance with remarkable success, its grave melancholy contrasting with the almost Beethovenian gruffness of the Scherzo. Yet Weber's difficulty in handling such emotional extremes without an instinctive command of Classical structures shows in the answering strain to this gesture, a flute waltz, obviating a formal trio, that might well have found its way into the *Aufforderung zum Tanze* but risks seeming trivial in this context. The finale begins by proposing a theme based on five contrasting elements, and the second theme is another graceful flute melody; the substance of the movement is an elaborate, contrapuntally inclined engagement between them. Though ingenious, it is not formally rigorous enough to chart much genuine emotional progress; but the work does reflect, in chamber music writing of great charm and ingenuity, the dramatically close conjunction of misery and consolation characterizing Weber's life at this period.

Of Weber's works for solo instrument and piano, the most important is the *Grand duo concertant* (with clarinet). It is no sonata (a description he abandoned during its composition) but a full-scale concert piece for two virtuosos. Weber skilfully devised themes that, while suiting each instrument, assume their full character only in combination. Even in the Andante, where the unique cantabile qualities of the clarinet demand acknowledgment, Weber gave the instruments equal status by answering the opening clarinet song with a powerful piano episode and then gradually reconciling and uniting the two. The finale, expectedly, is a tour de force of shared and rival brilliance; and, indeed, throughout the work Weber's imagination was most powerfully stimulated in

dramatizing the two instruments' relationship in a virtuoso context. The hastily assembled *Silvana Variations*
for clarinet and piano, while agreeably handled, are
unimportant; but Weber liked their theme well enough
to have used it before for variations in his *Six sonates
progressives* 'for piano with violin obbligato'. Also written in haste and unwillingly to commission, they are
nevertheless delightful pieces, mostly well within the
range of the amateur violinist and discovering within the
short space of each movement a number of charming
and original ideas.

VI Piano music

Behind the keyboard style of Weber lies the example
of the composer–virtuosos who had dominated music in
his formative years, among them C. P. E. Bach, Cramer,
and especially Hummel and Dussek. Weber's acute
feeling for keyboard colour probably found most
stimulus in Clementi, but much else, especially his adaptation of virtuoso methods to his own idiom, derives
from Cramer and Dussek, and from Prince Louis
Ferdinand, whom he greatly admired and whom he
saluted in the third volume of *Leyer und Schwert* by
making use of his themes. Characteristic devices which
he absorbed into his own virtuoso style include brilliant
passage-work in 3rds, widely ranging arpeggiated leaps,
the use of fast-running sequences, often chromatically
inflected, descending right down the keyboard, and
rapid staccato figuration carried over from the old
toccata but more sensational in expressive effect. He
also made particular use of quasi-vocal sonorities, with
a fondness for a singing melody over light repeated
chords resembling the guitar song and for the rapid

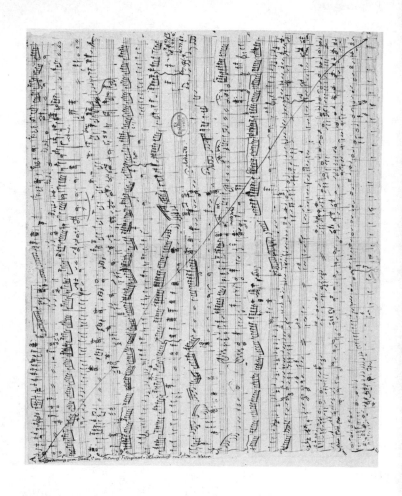

7. Autograph sketch for Weber's 'Aufforderung zum Tanze', composed 1819

64

chromatic scales and chromatically inflected flourishes that, as in the piano concertos, anticipate Chopin.

Liszt admired Weber enough to prepare editions of the piano music, including the sonatas; and it is possible that Chopin's affection for him took constructive form in the downward rush of semiquavers opening the 'Revolutionary' Study, which recalls the opening of Weber's First Piano Sonata. Written in 1812, the sonata is typical of Weber's music of this period, notably in the sense of drama taking precedence over formal demands in the first movement; in the work's greatest interest residing in its Adagio, whose ideas grow out of the sensitive use of piano sonorities; and in the scintillating virtuoso finale which Weber called 'L'infatigable'. The two piano sonatas of 1816 are less concerned with effect, developing the personal elements of the first while playing down the sensational ones. The first movement of no.2 was the most successful sonata movement he had then written, and though dramatic by nature is handled with greater command than the equivalent movement of no.3. The two Andantes are to different degrees in variation form; thereafter the works take on opposed characters, no.2 continuing with a somewhat over-regular Menuetto capriccioso and a finale of relaxed charm, no.3 proceeding straight to a finale combining three themes in a tour de force as much compositional as pianistic.

The Fourth Sonata is the most finished and unified of the four, a feature which is likely to derive from a programme concerning a sufferer from melancholy who breaks out in rage and insanity, seeks consolation, but finally collapses in exhaustion and death (Weber evidently gave Benedict this description). The music has a consistent emotional charge lacking in any of the previous sonatas, and its skilful use of a pervasive falling scalic figure suggests that with more time and energy Weber might have developed in the direction of the one-movement sonata.

Weber also wrote a number of individual virtuoso works for himself to play, of which by far the most famous is the *Aufforderung zum Tanze*. Even if Caroline had not communicated the details of the music's programme, which Weber gave her when he first played it to her, it would be simple to deduce the outline plot of a dancer approaching the lady of his choice, bearing her off into a brilliant waltz and finally taking courteous leave of her. But as is often the case with Weber, the value of the programme was strictly musical in stimulating a satisfying and in this instance simple formal plan. Through Berlioz's sympathetic scoring, made for use at his Paris performance of *Der Freischütz* with his own added recitatives in 1841, it has become widely known as an orchestral piece; but it is essentially a bravura keyboard work in Weber's most elegant manner, and at a stroke it placed the waltz in the Romantic repertory alongside the polonaise, nocturne and scherzo. The *Grande polonaise* was an earlier, shorter attempt at constructing a virtuoso piece without reference to sonata techniques, and was in 1808 the most elaborate structure Weber had attempted: no programme has been revealed, and it seems unlikely that one existed behind its self-sufficient cycle of modulations conceived in virtuoso terms. To the same period belongs the dazzling little *Momento capriccioso*, a pattering development of toccata technique that is closer to Mendelssohn than Clementi in spirit. The *Polacca bril-*

lante, splendidly effective in its own right, anticipates the Romantic concerto polonaise, while the *Rondo brillante*, another work of the fruitful summer of 1819, is nearer to the salon style of earlier days. Few of Weber's sets of variations can match the charm and virtuosity of these admirable concert solos; though he wrote eight, considerably the most ambitious and successful is that on *Schöne Minka*.

Weber also left three sets of piano duets. The *Six petites pièces faciles*, though very simple and early (1801), are engaging and well suited to beginner pianists. At the end of his life Weber arranged the march (no.5), with a different trio section, for wind band. The later set of six pieces (1809), written in connection with teaching, are slightly more demanding, and the mature set of eight more ambitious and correspondingly wider in expressive range and technique. All these charming duets are designed for and best appreciated in the home, not the concert platform.

Stylistic features

All Weber's most successful music is to a greater or lesser degree dramatically inspired. With the exception of that in the operas, it may have its stimulus in a miniature dramatic situation as in the songs, in the discovery of novel, sometimes theatrical ideas for the solo instruments of the concertos, in exploiting the virtuoso's personal relationship with his audience, or in conceiving and developing the materials of an extended movement either by relation to a programme or as dramatic contrasts. Although this can weaken the music with an inappropriate theatricality, it can also produce effects of striking originality. In developing forms which demonstrated that dramatic or narrative or pictorial elements could determine the actual structure of music, he deeply influenced Romantic music.

As an orchestrator, Weber was among the greatest in history. His discovery of new characteristics in instruments, especially his beloved clarinet and horn, led him to write for the orchestra as a group of individuals; although he could always write effective and powerful tuttis in classical proportions, he was at his most characteristic and most inventive when drawing dramatically appropriate music from the unexpected – but always apt and vivid – combination of instruments, perhaps in surprising registers. The overture to *Oberon* includes an example that greatly impressed Berlioz in its setting of violas and cellos above two clarinets in their lowest register; several times, including in *Der Freischütz* and *Euryanthe*, he effectively combined muted and unmuted strings; Caspar's piccolos in *Der Freischütz* have a diabolical glitter that arises from precise understanding of the instrument's register and the effect of its trill; his understanding of the horn ranges from the ideal use of hunting effects to the sinister blare in the Wolf's Glen and to the soft, evocative call that sounds all through *Oberon*. More concerned for a subtle individuality than for a rich blend of orchestral sound, he is in this respect an ancestor of Berlioz and Mahler rather than of Wagner; he was praised for his orchestration by Debussy.

Harmonically, Weber's music is rooted in Classical principles. As he matured he experimented more daringly with chromatic harmony, initially by using the diminished 7th, which at one point became something of a cliché in his music. Later he made greater use of 7th and 9th chords, often unconventionally approached; and by the time of *Euryanthe* he had developed a command of free-moving chromatic harmony that took him to the frontiers of tonality. The opening of Act 3 of *Euryanthe* for first and second violins, though in D minor, goes through a highly chromatic series of progressions, including a prominent use of the 'Tristan chord', that doubtless made Weber consider a key signature inappropriate (the prompt for this remarkable passage is the dramatic image of Euryanthe and Adolar wandering in a state of emotional disarray that prevents them from finding any security or mutual contact). In his piano music, too, the elaborate, highly inflected flourishes decorating the melody or a plain progression sometimes stray so far away from the basic tonality as to approach Chopin's similar, nearly atonal adventures. But he never lost his very strong rooting in tonality, and the harmonic ground structure of his operas shows the importance he gave to the traditional relationships between tonality and dramatic expression.

His melodies have a grace and elegance that is entirely personal, though often deceptively simple: a little-known Weber work can seem plain on the page, until performance or proper aural imagination shows how acutely a simple tune over a simple harmonic pattern has been judged and deftly given a particular atmosphere by an original touch in phrasing or harmonic inflection. Frequently, especially in his fast music, he made use of unprepared notes outside the accompanying chord, giving the music a peculiar tension or a subtly passionate quality. He was fond, perhaps over-fond, of dotted rhythms; and a weakness which he was slow to overcome was for allowing his tunes to flag in their second strain, often ending in a limp cadence. It was as he developed greater independence from the strict dictates of Classical harmony that he found a freer, more personal and stronger melodic voice. Rhythmically robust when the situation demanded, he could also write music that, by cunning alternation of expected lightly stressed beats over a simple harmonic structure with startling cross-rhythms and harmonic swerves, gives a sense of pace which not even Mendelssohn and Berlioz were to excel.

Weber's role as a father figure of Romanticism was acknowledged by all those who succeeded him in the movement; in modern times, homage has been paid him by composers as diverse as Debussy, Stravinsky and

Hindemith. At a time of receptivity to new ideas and of heightened emotional responses, he provided music that spoke of a new sensibility and that opened up connections between different arts and different activities. Thus, he aroused the interest of a politically conscious nation, as well as winning the admiration of writers and painters. Although he died before he could explore fully all the ideas he had opened up, he left a large body of work whose value is being more fully appreciated.

WORKS

Editions: *C. M. von Weber: Musikalische Werke: erste kritische Gesamtausgabe*, ed. H. J. Moser; ii/1, ed. A. Lorenz (Augsburg, 1926), ii/2, ed. W. Kaehler (Augsburg, 1928), ii/3, ed. L. K. Mayer (Brunswick, 1939) [WG]
Reliquienschrein des Meisters Carl Maria von Weber, ed. L. Hirschberg (Berlin, 1927) [HR]

J – no. in *Jähns (1871)*

Numbers in the right-hand column denote references in the text.

OPERAS AND SINGSPIELS
(all publications in full score unless otherwise stated)

40–52, 217

J	Title and genre	Libretto	Date	First performance	Publication/MS: Remarks	
Anh.6	Die Macht der Liebe und des Weins, Singspiel	—	1798	—	lost	2, 43
Anh.1	Das Waldmädchen (Das stumme Waldmädchen; Das Mädchen im Spessartwalde), Romantic-comic opera	2, C. von Steinsberg	1800	Freiberg (Saxony), 24 Nov 1800	frags, WG ii/1	2, 5, 43, 44
8	Peter Schmoll und seine Nachbarn, opera	2, J. Türk[e], after C. G. Cramer	1801–2	Augsburg, ?March 1803	WG ii/1; dialogue lost, ov. rev. 1807 as J54	2, 3, 43–4, 58
44–6	Rübezahl, opera	3, J. G. Rhode	1804–5	—	WG ii/2; 3 nos. survive	4, 58
87	Silvana, Romantic opera	3, F. C. Hiemer, after Das Waldmädchen text	1808–10	Frankfurt am Main, 16 Sept 1810	WG ii/2; vocal score without ensemble scenes (Berlin, 1812), complete vocal score (Berlin, 1828)	5, 8, 12, 43, 44–5
106	Abu Hassan, Singspiel	1, Hiemer, after The 1001 Nights	1810–11	Munich, Residenz, 4 June 1811	ed. W. W. Götig (Offenbach, 1925), vocal score (Bonn, 1819)	7, 22, 45, 47
277	Der Freischütz, Romantic opera	3, F. Kind, after J. A. Apel and F. Laun: Gespensterbuch	1817–21	Berlin, Schauspielhaus, 18 June 1821	(Berlin, 1843), vocal score (Berlin, 1822)	7, 16, 17, 18, 19, 21, 22, 23, 24, 34, 39, 41, 44, 46, 47–8, 49, 51, 52, 66, 69, 92, 109, 134, 165, 214
Anh.5	Die drei Pintos, comic opera	3, T. Hell, after C. Seidel	1820–21	Leipzig, Neues Stadt-Theater, 20 Jan 1888	inc., completed by G. Mahler; text rev. C. Weber and Mahler (Leipzig, 1888)	18, 19, 51
291	Euryanthe, grand heroic-Romantic opera	3, H. von Chezy	1822–3	Vienna, Kärntnertor., 25 Oct 1823	ed. E. Rudorff (Berlin, 1866), vocal score (Vienna, 1824)	19, 20, 21, 22–3, 25, 28, 39, 40, 41–2, 44, 45, 49–50, 51, 60, 69, 230
306	Oberon, Romantic opera	3, J. R. Planché, after C. M. Wieland	1825–6	London, Covent Garden, 12 April 1826	(Berlin, 1872), vocal score (London, 1826)	23, 24, 25, 26, 27, 28, 35, 37, 41, 42–3, 45, 50–51, 52, 68–9, 214, 239, 244

(unless otherwise stated, all with orch and published in full score)

J

J		
75	Turandot, Prinzessin von China (Gozzi, trans. Schiller), incidental music, Stuttgart, Sept 1809 (Berlin, 1818) [ov., 6 inst nos.]	13, 52
77–8	2 numbers for Haydn pasticcio Der Freibrief, 1809, vocal score ed. F. W. Jähns (Berlin, 1839): Was ich da tu, rondo alla polacca, T [orig. in *Album drei neuester Original-Compositionen*, vocal score (Paris, c1837)]; Dich an dies Herz, S, T	5, 52
110–13	4 songs in Der arme Minnesinger (Kotzebue), gui acc., 1811: Über die Berge (Berlin, ?1812); Rase, Sturmwind, blase, HR; Lass mich schlummern (Berlin, ?1812); Umringt von muterfüllten Heere, with male vv (Berlin, ?1812)	
178	Ah, se Edmondo, scena for Méhul's opera Héléna, 1815, vocal score and parts (Berlin, 1826)	
183–4	2 songs for A. Fischer's Singspiel Der travestirte Aeneas, 1815: Mein Weib ist capores, Bar (Berlin, c1838); Frau Lieserl juhel, S, B, HR [arr. for orch as J185]	
186–7	2 songs in Lieb und Versöhnen (F. W. Gubitz), 1815 (Berlin, c1838)	
189	Was stürmet, ballad in Gordon und Montrose (G. Reinbeck), Bar, harp, 1815 (Berlin, 1818)	
194	Arietta in Das Sternenmädchen im Maidlinger Walde (L. Huber and F. Kauer), 1816, ?D-Bds [text lost]	
195	Ein König einst gefangen sass, romance in Diana von Poitiers (I. Castelli), gui acc., 1816 (Berlin, 1822)	
214	König Yngurd (A. Müllner), incidental music, Dresden, Hoftheater, 14 April 1817, HR [10 inst nos., 1 song, Mez unacc.]	
220	Donna Diana (A. Moreto), incidental music, Dresden, Hoftheater, 1817, HR [6 nos.]	
222	Hold ist der Zyanenkranz, song in Der Weinberg an der Elbe (F. Kind), solo vv, chorus, 1817, *Malerische Schauspiele* (Leipzig, 1817), suppl.	16
223	Leise weht es, romance in Das Nachtlager von Granada (Kind), 1818, ed. J. Otto (Dresden, 1832)	
225	Sei gegrüsst, Frau Sonne, mir, song in Die drei Wahrzeichen (F. von Holbein), T, B, 1818, ed. J. Otto (Dresden, 1832)	
227	In Provence, dance and song with Andantino in Das Haus Anglade (T. Hell), T, chorus, 1818, HR [dance and song ?spurious]	
237	Heinrich IV, König von Frankreich (E. Gehe), incidental music, Dresden, Hoftheater, 6 June 1818, HR [8 inst nos.]	52
239	Was sag ich?, scena and aria for Cherubini's opera Lodoiska, S, 1818, vocal score and parts (Berlin, 1824)	
240	Heil dir, Sappho, chorus in Sappho (Grillparzer), SSB, wind, perc, 1818, HR	
243	Ein Mädchen ging, song in Der Abend am Waldbrunnen (Kind), pf/gui acc., 1818 (Berlin, 1819)	
246	Lieb um Liebe (A. Rublack), incidental music, 1818, HR [6 vocal and inst nos.]	
273	Agnus Dei in Carlo (G. von Blankensee), SSA, wind, 1820, HR	52
276	Der Leuchtturm (E. von Houwald), incidental music, Dresden, Hoftheater, 26 April 1820, HR [4 harp nos.]	
279	Preciosa (P. A. Wolff), incidental music, 1820, Berlin, Hoftheater, 14 March 1821 (Berlin, c1842), WG ii/3, vocal score (Berlin, c1823) [ov., 11 nos., solo v, chorus orch]	8, 17, 18, 21, 31, 51–2
280	Sagt, woher, song in The Merchant of Venice (Shakespeare), S, S, A, SSA, gui, 1821, HR	
289	Den Sachsensohn vermählet heute (L. Robert), incidental music, 1822, HR [ov., 5 choral nos.]	
305	Doch welche Töne, arioso and recit for Spontini's opera Olympie, B, S, 1825, HR	24–5

CONCERT ARIAS

(published in full score unless otherwise stated)

J		
93	Il momento s'avvicina, recit and rondo, S, orch, 1810, vocal score and parts (Offenbach, 1810)	
121	Misera me!, scena and aria for Atalia, S, orch, 1811 (Berlin, 1818)	
126	Qual altro attendi, scena and aria, T, chorus, orch, 1812, HR	
142	Signor, se padre sei, scena and aria for Ines de Castro, T, 2 choruses, orch, 1812 (Berlin, 1824)	
181	Non paventar mia vita, scena and aria for Ines de Castro, S, orch, 1815, vocal score and parts (Berlin, 1818)	

SACRED CHORAL
(all published in full score)

No.	refs	
Anh.8	52–3	Mass (Grosse Jugendmesse), E♭, S, A, T, B, SATB, orch, org, 1802, ed. C. Schneider (Augsburg, 1926)
224	2, 52	Missa sancta no.1, E♭, S, A, T, B, SATB, orch, 1817–18 (Vienna, 1844)
226	17, 52–3	Gloria et honore, off, S, SATB, orch, 1818, ?D-Bds [for Missa sancta no.1]
250		In die solemnitatis, off, S, SATB, orch, 1818, HR [for Missa sancta no.2]
251	17, 53	Missa sancta no.2 (Jubelmesse), G, S, A, T, B, SATB, orch. 1818–19 (Vienna, 1835)

CANTATAS
(published in full score unless otherwise stated)

No.	refs	
58	53–4	Der erste Ton (F. Rochlitz), reciter, SATB, orch, 1808, rev. 1810, parts (Bonn, 1811)
154	53	In seiner Ordnung schafft der Herr (Rochlitz), hymn, S, A, T, B, SATB, orch, 1812 (Berlin, 1817)
190	13, 14, 31, 54, 58	Kampf und Sieg (J. G. Wohlbrück), S, A, T, B, SATB, orch, 1815, vocal score (Berlin, 1817)
221	54	L'accoglienza (T. Celani), 3 S, T, 2 B, SATB, orch, 1817
244	54, 58	Jubel-Kantate (Kind), S, A, T, B, SATB, orch, 1818 (Berlin, 1819)
283		Du, bekränzend unsre Laren (Kind), 2 S, T, B, SATB, fl, pf, 1821, ed. C. Banck (Leipzig, c1879)

OTHER CHORAL ACCOMPANIED

No.	refs	
—	53	An die Hoffnung (anon.), canon, 3vv, pf ad lib, 1802, ed. in Kroll (1934) [attrib. by E. Kroll]
37		Grablied (anon.), TTB, 9 wind, 1803, ed. F. W. Jähns (Berlin, 1840)
66		Die Lethe des Lebens (J. Baggesen), B solo, SATB, pf, 1809 (Berlin, 1819)
116		Trauer-Musik (anon.), Bar, SATB, 10 wind, 1811, HR
131		Prost Neujahr! (?Weber), canon, 34vv, 74 tpt, 1811, HR
133		Lenz erwacht, und Nachtigallen (anon.), S, T, B, STB, pf. 1812, copy, Bds

No.	refs	
135		Schwäbisches Tanzlied (S. F. Sauter), STTBB, pf, 1812 (Berlin, c1813)
136		Heisse, stille Liebe schwebet (anon.), STTB, pf, 1812 (Berlin, 1812)
139	17, 52–3	Kriegs-Eid (H. J. von Collin), unison male vv, 2 tpt, 3 hn, bn, trbn, 1812, HR
165		Lebenslied am Geburtstage (F. W. Gubitz), 4 male vv, pf, 1814 (Berlin, 1819)
218		Zwei Kränze zum Annen-Tage (Kind), 4 male vv, pf, 1817 (Berlin, 1819)
228	17, 53	Schöne Ahnung ist erglommen (Kind), 4 male vv, pf, 1818 (Berlin, 1819) [also with texts Schmückt das Haus (H. Illaire), Singet dem Gesang zu Ehren (anon.) and others]
241	53–4	Natur und Liebe (Kind), SSTTBB, pf, 1818 (Berlin, 1823) [later with text Freundschaft und Liebe (F. Herklots)]
271	5, 6, 53–4	Music, and song Du hohe Rautenzweig [melody of God Save the King], for ceremonial prologue (T. Hell), SATB, 6 wind, 1819, HR
Anh.3	53	Deo rosa (Hell), 4 male vv, pf, 1821, inc., HR
290	13, 14, 31, 54, 58	Wo nehm ich Blumen her (Hell), small cantata, STB, pf, 1823, HR
293	54	Reiterlied (E. Reiniger), 4 male vv, pf ad lib, 1825, HR
294	54, 58	Schützenweihe (A. Oertel), 4 male vv, pf ad lib, 1825, Liederbuch für deutsche Krieger und deutsches Volk, ed. C. Weitershausen (Darmstadt, 1830, 2/1837)
[307]	53	Zu den Fluren des heimischen Herdes, SATTB, orch, 1826 (Leipzig, 1858) [arr. of March, wind, J307]

OTHER CHORAL UNACCOMPANIED

No.	refs	
35		Mädchen, ach meide Männerschmeichelei'n (Breiting), canon, 3vv, 1802 (Augsburg, 1811)
—		Wenn du in Armen der Liebe (anon.), canon, 1802
36		Ein Gärtchen und ein Häuschen drin (anon.), STB, 1803, copy, ?D-Bds
69		Chorlied, SATB, 1809, HR [no text]
89		Die Sonate soll ich spielen (Weber), canon, 3vv, 1810, ed. in Jähns (1871), 109
90		Canons zu zwei sind nicht drei (anon.), canon, 3vv, 1810, Der musikalische Hausfreund, i, ed. F. S. Gassner (Mainz, 1822)

Part-songs (continued):

No.		Description
95		Leck mich im Angesicht (anon.), canon, 3vv, 1810, ed. in Jähns (1871), 116
132		Das Turnierbankett (F. W. Bornemann), 2 T, B, TTBB, TTBB, 1812 (Berlin, 1823)
164		Zu dem Reich der Töne schweben (F. W. Gubitz), canon, 4vv, 1814, ed. in F. W. Gubitz: *Volkskalender* (Berlin, 1862)
167		Scheiden und leiden ist einerlei (anon.), canon, 4vv, 1814, ed. in Jähns (1871), 182
—	54-5	Leise kömt der Mond gezogen (anon.), canon, 4vv, 1814, HR
168-73		Leyer und Schwert (T. Körner), ii, TTBB, 1814 (Berlin, 1816): 2 Lützows wilde Jagd, 6 Schwertlied, 4 Männer und Buben, 5 Trinklied vor der Schlacht, 1 Reiterlied, 3 Gebet vor der Schlacht
180		Drei Knäbchen lieblich austaffiret (Weber), burlesque on Mozart's Die Zauberflöte, TTB, 1815, ed. in L. Nohl: *Musiker-Briefe* (Leipzig, 1867), 281ff
193		Weil Maria Töne hext (Gubitz), canon, 3vv, 1816, HR
249		Ei, ei, wie scheint der Mond so hell (trad.), TTB, 1818 (Berlin, 1822)
261-3	53	3 partsongs (K. Kannegiesser), 4 male vv, 1819 (Berlin, 1823): Gute Nacht, Freiheitslied, Ermunterung
272		Double canon, 4vv, 1819, HR [no text]
284-5		2 partsongs, 4 male vv (Berlin, 1823): Husarenlied (A. vom Thale), 1821, Schlummerlied (I. Castelli), 1822

DUETS

No.		Description
107		Se il mio ben (anon.), 2 A, cl, 2 hn, str, 1811, acc. pf (Berlin, 1815)
123		Mille volte (anon.), 2 S, pf, 1811 (Berlin, 1815)
125		Va, ti consola (anon.), 2 S, pf, 1811 (Berlin, 1815)
208		Abschied: O Berlin, ich muss dich lassen (trad.), 2vv, pf, 1817 (Leipzig, 1818)
209		Quodlibet: So geht es in Schnützelputz-Häusel (trad.), 2vv, pf, 1817 (Leipzig, 1818)
210	3, 5, 55-7	Mailied: Tra, ri, ro! (trad.), 2vv, pf, 1817 (Berlin, 1822)

SONGS

(with pf accompaniment unless otherwise stated)

No.	Description
—	Strafpredigt über die französische Musik und Übersetzungswut (?Weber), 1801, HR [no text]
27	Die Kerze (anon.), 1802, copy, ?*D-Bds*
28	Umsonst entsagt ich der lockende Liebe (anon.), 1802 (Hamburg, n.d.)
38	Entfliehet schnell von mir (F. von Seida), 1803, HR
41	Ich sah sie hingesunken (Swoboda), 1804, HR
42	Wiedersehen (Wallner), 1804 (Berlin, 1815)
48	Ich denke dein (Matthison), 1806 (Berlin, 1819)
52	Liebeszauber (J. H. Bürger), gui acc., 1807 (Augsburg, 1811)
57	Er an Sie (F. Lehr), 1808 (Bonn, 1819)
60	Komisches musikalisches Sendschreiben (Weber), bc acc., 1808, ed. in M. M. von Weber (1864), 146ff
62	Meine Farben (Lehr), 1808 (Berlin, 1812)
63	Klage (C. Müchler), 1808 (Bonn, 1819)
65	Serenade (C. Baggesen), 1809, *Morgenblatt* (Stuttgart, 8 Jan 1810)
67	Das Röschen (Müchler), 1809 (Bonn, 1819)
68	Was zieht zu deinem Zauberkreise (Müchler), 1809 (Bonn, 1819)
70	Rhapsodie (Die Blume) (F. Haug), 1809 (Berlin, 1812) [orig. in *Morgenblatt* (Stuttgart, 1810)]
71	Romanze (Die Ruinen) (G. Reinbeck), 1809, *G. Reinbeck: Winterblüten*, i (Leipzig, 1810)
72	Sanftes Licht, weiche nicht (An den Mond) (Reinbeck), gui acc., 1809 (Augsburg, 1811)
73	Meine Lieder, meine Sänge (W. von Löwenstein-Wertheim), 1809 (Bonn, 1819)
74	Der kleine Fritz an seine jungen Freunde (trad.), 1809 (Bonn, 1819)
80	Trinklied (Lehr), 1809, HR
88	Sicché t'inganni (anon.), canzonetta, pf/harp acc., 1810 (Leipzig, 1853)
91	Die Schäferstunde (Damon und Chloe) (F. C. Hiemer), gui acc., 1810 (Augsburg, 1811)
92	Das neue Lied (Herder), 1810, HR [no text]
96	Wiegenlied (Hiemer), gui acc., 1810 (Augsburg, 1811)
97	Die Zeit (J. L. Stoll), gui acc., 1810 (Augsburg, 1811)
105	Des Künstlers Abschied (A. von Dusch), gui/pf acc., 1810 (Berlin, ?1819)
108	Ah, dove siete (anon.), canzonetta, gui/pf acc., 1811 (Prague, 1814)

57

J	Work	Page
117	Maienblümlein, so schön (A. Eckschläger), 1811 (Berlin, 1812)	
120	Ch'io mai vi possa (anon.), canzonetta, gui/pf acc., 1811 (Prague, 1814)	
124	Ninfe se liete (anon.), canzonetta, gui/pf acc., 1811 (Prague, 1814)	
129	Romanze (Wiedersehn) (Duke Leopold August of Gotha), 1812, *Polyhymnia auf 1826*, ed. F. Kind and H. Marschner (Leipzig, 1826)	
130	Sonett (C. Streckfuss), 1812 (Berlin, 1812)	
134	Lebensansicht (anon.), 1812 (Berlin, 1819)	
137	Bettlerlied (trad.), gui/pf acc., 1812 (Berlin, ?1812)	
140	Liebeglühen (F. W. Gubitz), gui/pf acc., 1812 (Berlin, ?1812)	
156	Sind es Schmerzen (Tieck), 1813 (Berlin, 1815)	
157	Unbefangenheit (anon.), 1813 (Berlin, 1815)	
159	Reigen (J. H. Voss), 1813 (Berlin, 1815)	
160	Minnelied (Voss), 1813 (Berlin, 1815)	56
161	Es stürmt auf der Flur (F. Rochlitz), 1814 (Berlin, 1815)	
166	Gebet um die Geliebte (Gubitz), 1814 (Berlin, 1818)	
174–7	Leyer und Schwert (T. Körner), i, 1814 (Berlin, 1815): Gebet während der Schlacht, Abschied vom Leben, Trost, Mein Vaterland	12, 54, 55
192	Der Jüngling und die Şŗröde (Gubitz), 1816 (Berlin, 1818)	
196	Mein Verlangen (F. Förster), 1816 (Berlin, 1818)	
197	Die gefangenen Sänger (M. von Schenkendorf), 1816 (Berlin, 1818)	
198	Die freien Sänger (Förster), 1816 (Berlin, 1818)	
200–03	Die Temperamente beim Verluste der Geliebten (Gubitz), cycle, 1816 (Berlin, 1817): Der Leichtmütige, Der Schwermütige, Der Liebeswütige, Der Gleichmütige	15, 57
205	Bei der Musik des Prinzen Louis Ferdinand von Preussen: Düstre Harmonien hör ich klingen, Leyer und Schwert (Körner), iii, 1816 (Berlin, 1817)	63
211	Alte Weiber (F. Nicolai), 1817 (Leipzig, 1818)	
212	Liebeslied (trad.), 1817 (Leipzig, 1818)	
213	Wunsch und Entsagung (I. Castelli), 1817 (Berlin, 1819)	
217	Das Veilchen im Tale (F. Kind), 1817 (Berlin, 1819)	
229	Lied der Hirtin (Kind), 1818 (Berlin, ?1819)	
230	Gelahrtheit (M. Opitz), 1818 (Berlin, 1822)	
231	Weine, weine (trad.), 1818 (Leipzig, 1818)	
232	Die fromme Magd (B. Ringwald), 1818 (Leipzig, 1818)	
233	Wenn ich ein Vöglein wär (trad.), 1818 (Leipzig, 1818)	
234	Mein Schatzerl is hübsch (trad.), 1818 (Berlin, 1822)	
235	Heimlicher Liebe Pein (trad.), 1818 (Berlin, 1822)	
238	Rosen im Haare (F. L. B. Breuer, after Hafiz), 1818 (Berlin, 1819)	
255	Abendsegen (trad.), 1819 (Berlin, 1822)	
256	Triolett (Förster), 1819 (Berlin, ?1819)	
257	Liebesgruss aus der Ferne (trad.), 1819 (Berlin, 1822)	
258	Herzchen, mein Schätzchen (trad.), 1819 (Berlin, 1822)	
267	Das Mädchen an das erste Schneeglöckchen (F. von Gerstenbergk), 1819 (Berlin, ?1819)	
269	Sehnsucht (Weihnachtslied) (K. Kannegiesser), 1818 (Berlin, 1823)	
270	Elfenlied (Kannegiesser), 1819 (Berlin, 1823)	
274	Schmerz (G. von Blankensee), 1820 (Berlin, 1823)	23
275	An Sie (Wargentin), 1820 (Berlin, 1823)	
278	Der Sänger und der Maler (anon.), 1820 (Berlin, 1823)	28
281	Lied von Clotilde (C. von Nostiz und Jänkendorf), 1821 (Berlin, 1823)	
286	Das Licht im Tale (Kind), 1822, *C. W. Beckers Taschenbuch*, ed. F. Kind (Berlin, 1823)	
—	Vatergruss (Weber), 1823, HR [no text]	
292	Elle était simple et gentilette (F. de Cussy), romance, 1824 (Paris, ?1824)	
308	Gesang der Nurmahal: From Chindara's warbling fount I come (T. Moore: Lalla Rookh), 1826, ed. F. W. Jähns (Berlin, 1869) [pf part reconstructed by Moscheles]	58–61, 217
Anh.67–70	4 solfèges, 1818, HR	

ORCHESTRAL AND WIND

J	Work	Page
47	Romanza siciliana, g, fl solo, 1805 (Berlin, 1839)	
47a	Tusch, 20 tpt, 1806, ed. in Jähns (1871)	
49	6 variations on 'A Schüsserl und a Reind'rl', C, va solo, 1806, HR	
50	Symphony no.1, C, 1807 (Offenbach, 1812)	4, 58

J.	Work	Pages
51	Symphony no.2, C, 1807 (Berlin, 1839)	4, 58
54	Grande ouverture à plusieurs instruments, 1807 (Augsburg, ?1807) [rev. ov. to Peter Schmoll und seine Nachbarn 8]	58
64	Grand pot-pourri, vc solo, 1808 (Bonn, 1822)	
79	Andante e Rondo ungarese, c, va solo, 1809, ed. G. Schünemann (Mainz, 1938) [rev. as J158]	59
94	Variations, F, vc solo, 1810 (Leipzig, c1814)	
98	Piano Concerto no.1, C, 1810 (Offenbach, 1812)	6, 7, 60-61
109	Clarinet Concertino, E♭, 1811 (Leipzig, 1814)	7, 59-60
114	Clarinet Concerto no.1, f, 1811 (Berlin, 1822)	7, 60
115	Adagio und Rondo, F, harmonichord/harmonium, 1811 (Leipzig, 1861)	59
118	Clarinet Concerto no.2, E♭, 1811 (Berlin, 1823)	7, 60
122	Der Beherrscher der Geister, ov., 1811 (Leipzig, 1812)	58
127	Bassoon Concerto, F, 1811, rev. 1822 (Berlin, 1824)	7, 60
149	Waltz, fl, 2 cl, 2 bn, tpt, 2 bn, 1812, HR	
155	Piano Concerto no.2, E♭, 1812 (Berlin, 1814)	10, 33, 60-61
158	Andante e Rondo ungarese, c, bn solo, 1813 (Berlin, c1816) [rev. of J79]	59
185	Deutscher (Original-Walzer), D, 1815 (Berlin, c1842-3) [arr. of J184, song for Singspiel Der travestirte Aeneas]	
188	Horn Concertino, e, 1806, rev. 1815 (Leipzig, 1818) [1st version lost]	4, 13, 59
191	Tedesco, D, 1816, HR	
245	Jubel-Ouvertüre, E, 1818 (Berlin, 1819)	58, 220
282	Konzertstück, f, pf solo, 1821 (Leipzig, 1823)	18, 19, 33, 59, 61
288	Marcia vivace, 10 tpt, 1822, HR	
307	March, wind, 1826 (Leipzig, 1853) [rev. of pf duet J13, with new trio; also arr. for chorus, orch]	28, 67
—	Oboe Concertino [not authenticated; probably spurious]	

CHAMBER

J.	Work	Pages
61	Neuf variations sur un air norvégien, vn, pf, 1808 (Berlin, 1812)	12, 61-3
76	Piano Quartet B♭, 1809 (Bonn, 1811)	5
99-104	Six sonates progressives, vn, pf, 1810 (Bonn, 1811): F, G, d, E♭, F, C	5, 61
119	Melody, F, cl, 1811, with pf acc. by F. W. Jähns in Album deutscher Komponisten, xii, ed. H. Mohr (Berlin, 1872)	7, 63
128	Seven variations on a theme from Silvana, cl, pf, 1811 (Berlin, 1814)	63
182	Clarinet Quintet, B♭, 1815 (Berlin, 1816)	61
204	Grand duo concertant, cl, pf, 1815-16 (Berlin, 1817)	13, 15, 62-3
207	Divertimento assai facile, gui, pf, 1816 (Berlin, 1818)	
259	Trio, fl, vc, pf, 1819 (Berlin, 1820)	17, 61-2

Spurious: Introduction, Theme and Variations, cl, str qt = Clarinet Quintet, B♭, op.32, by J. Küffner

PIANO

(for 2 hands unless otherwise stated)

J.	Work	Pages
1-6	Sechs Fughetten, 1798 (Salzburg, 1798)	3, 32, 63-7
7	Six variations on an original theme, 1800 (Munich, 1800) [lithography (inaccurate) by Weber]	2, 53
9-14	Six petites pièces faciles, 1801 (Augsburg, 1803)	
15-26	Douze allemandes, 1801 (Augsburg, 1801) [nos.11-12 for 4 hands]	28, 67
29-34	Sechs Ecossaisen, 1802 (Hamburg, 1802)	
40	Huit variations sur l'air de ballet de Castor et Pollux [Vogler's opera], 1804 (Vienna, 1804)	3
43	Six variations sur l'air de Naga 'Woher mag dies wohl kommen?' [from Vogler's opera Samori], pf (vn, vc ad lib), 1804 (Vienna, 1804)	3
53	Sept variations sur l'air 'Vien quà, Dorina bella', 1807 (Augsburg, 1811)	
55	Thème original varié, 1808 (Offenbach, 1810) [7 variations]	5, 66
56	Momento capriccioso, B♭, 1808 (Augsburg, 1811)	5, 66
59	Grande polonaise, E♭, 1808 (Bonn, 1810)	5, 67
81-6	Six pièces, 4 hands, 1809 (Augsburg, c1810)	
138	Sonata no.1, C, 1812 (Berlin, 1812)	10, 32, 65
141	Seven variations on 'A peine au sortir de l'enfance' from Méhul's opera Joseph, 1812 (Leipzig, 1812)	
143-8	Sechs Favorit-Walzer der Kaiserin von Frankreich, Marie Louise, 1812 (Leipzig, 1812)	5
179	Air russe (Schöne Minka), 9 variations, 1815 (Berlin, 1815)	12, 67
199	Sonata no.2, A♭, 1816 (Berlin, 1816)	15, 65
206	Sonata no.3, d, 1816 (Berlin, 1817)	15, 65
219	Sieben Variationen über ein Zigeunerlied, 1817 (Berlin, 1819)	

236, 242, 248, 253–4, 264–6	Huit pièces, 4 hands, 1818–19 (Berlin, 1818–19)	67

236, 242, 248, 253–4, 264–6 — Huit pièces, 4 hands, 1818–19 (Berlin, 1818–19) — 67

252 Rondo brillante (La gaité), E♭, 1819 (Berlin, 1819) — 17, 66

260 Aufforderung zum Tanze: rondo brillant, D♭, 1819 (Berlin, 1821) — 17, 62, *64*, 66

268 Polacca brillante (L'hilarité), E, 1819 (Berlin, 1819) — 17, 66

287 Sonata no.4, e, 1819–22 (Berlin, 1823) — 17, 21, 65–6

EDITIONS/ARRANGEMENTS

39 G. J. Vogler: Samori, arr. vocal score, 1803 (Vienna, ?1804) — 3

150–53 Duke Leopold August of Gotha: 4 songs, acc. arr. wind, 1812, ?*D-Bds*

162 J. Weigl: Ein jeder Geck, duet in Fischer's Singspiel Die Verwandlungen, 2 S, arr. and orchd, 1814

163 Ihr holden Blumen, anon. arietta in Fischer's Singspiel Die Verwandlungen, S, arr. and orchd, 1814

215 F. Paer: Von dir entfernt, recit and cavatina, S, orchd for Méhul's Hélèna, 1817

216 S. Nasolini: Ja, Liebe, recit and duet, S, T, orchd for Méhul's Hélèna, 1817

247 God Save the King, arr. TTBB, ?1818, HR

295–304 Zehn schottische Nationalgesänge, acc. arr. fl, vn, vc, pf, 1825 (Leipzig, 1826): The soothing shades of gloaming (T. Pringle), The Troubadour (Scott), O poortith cauld (Burns), Bonny Dundee (Burns), Yes, thou may'st walk (J. Richardson), A soldier am I (W. Smyth), John Anderson, my jo' (Burns), O my love's like the red, red rose (Burns), Robin is my joy (D. Vedder), Whar' hae ye been a' day (H. Macnell) — 24, 55

WRITINGS

T. Hell, ed.: *Hinterlassene Schriften von Carl Maria von Weber* (Dresden and Leipzig, 1828, 2/1850) — 13, 17, 29–32, 37, 43

M. M. von Weber, ed.: *Carl Maria von Weber: ein Lebensbild*, iii (Leipzig, 1866) [selected writings]

G. Kaiser, ed.: *Sämtliche Schriften von Carl Maria von Weber: Kritische Ausgabe* (Berlin and Leipzig, 1908) — 54

K. Laux, ed.: *Carl Maria von Weber: Kunstansichten* (Leipzig, 1969, 2/1975) [selected writings]

J. Warrack, ed.: *Carl Maria von Weber: Writings on Music*, trans. M. Cooper (Cambridge, 1981)

BIBLIOGRAPHY
SOURCE MATERIAL, CATALOGUES, BIBLIOGRAPHIES, ICONOGRAPHY

F. W. Jähns: *Weberiana, D-Bds* [collected and catalogued MSS, printed books and music, documents etc]

——: *Carl Maria von Weber in seinen Werken: chronologisch-thematisches Verzeichniss seiner sämmtlichen Compositionen* (Berlin, 1871/*R*1967)

F. Rapp: *Ein unbekanntes Bildnis Carl Maria von Webers* (Stuttgart, 1937)

R. and L. Stebbins: *Enchanted Wanderer* (New York, 1940) [with extensive bibliography]

H. Dünnebeil: *Carl Maria von Weber: Verzeichnis seiner Kompositionen* (Berlin, 1942, 2/1947)

——: *Schrifttum über Carl Maria von Weber* (Berlin, 1947, 4/1957)

CORRESPONDENCE

Letters to G. Weber, *Caecilia*, iv (1826), 302; vii (1828), 20; xv (1833), 30

14 letters to F. von Mosel, 1 to Dr Jungh, *Wiener allgemeine Musikzeitung*, vi (1846), 473

C. von Weber, ed.: *Reise-Briefe von Carl Maria von Weber an seine Gattin Carolina* (Leipzig, 1886)

E. Rudorff, ed.: *Briefe von Carl Maria von Weber an Hinrich Lichtenstein* (Brunswick, 1900) [repr. from *Westermanns illustrierte deutsche Monatshefte*, lxxxvii (1899–1900), 16, 161, 367]

G. Kaiser, ed.: *Webers Briefe an den Grafen Karl von Brühl* (Leipzig, 1911)

O. Hellinghaus, ed.: *Karl Maria von Weber: seine Persönlichkeit in seinen Briefen und Tagebüchern und in Aufzeichnungen seiner Zeitgenossen* (Freiburg, 1924)

J. Kapp: 'Webers Aufenthalt in Berlin im August 1814 nach unveröffentlichten Briefen an seinen Braut', *Die Musik*, xviii (1925–6), 641

L. Hirschberg, ed.: *Siebenundsiebzig bisher ungedruckte Briefe Carl Maria von Webers* (Hildburghausen, 1926)

G. Kinsky: 'Ungedruckte Briefe Carl Maria v. Webers', *ZfM*, Jg.93 (1926), 335, 408, 482

W. Virneisel: 'Aus dem Berliner Freundeskreise Webers', *Carl Maria von Weber: eine Gedenkschrift*, ed. G. Hausswald (Dresden, 1951) [31 letters to F. Koch]

L. Nohl, ed.: *Mosaik: Denksteine* (Leipzig, n.d.), 63–93

79

Weber

PERSONAL REMINISCENCES

L. Rellstab: 'Weber', *Caecilia*, vii (1828), 1

G. Hogarth: *Musical History, Biography and Criticism* (London, 1835, 2/1838)

V. Tomášek: 'Selbstbiographie', *Jb für Libussa*, v (1846), 321–76

A. Schmidt: *Denksteine* (Vienna, 1848) [incl. part of Gänsbacher's autobiography]

K. von Holtei: 'Zur Erinnerung an Weber', *Wiener Modespiegel*, i (1853), 52, 70, 87; repr. in *Charpie* (Breslau, 1866), 325

J. Duesburg: 'Une visite à Weber en 1825', *Revue et gazette musicale*, xxii (1854), 231; also in *Modes parisiennes* (1854), 1746

J. C. Lobe: 'Gespräche mit Carl Maria von Weber', *Fliegende Blätter für Musik*, i (Leipzig, 1855), 27, 110; repr. in *Consonanzen und Dissonanzen* (Leipzig, 1869), 122ff

H. von Chezy: *Unvergessenes*, ed. B. Borngräber (Leipzig, 1858)

L. Spohr: *Selbstbiographie* (Kassel and Göttingen, 1860–61; Eng. trans., 1865/R1969, 2/1878)

I. Castelli: *Memoiren meines Lebens* (Vienna, 1861); ed. J. Bindtner (Munich, 1913)

W. von Chezy: *Erinnerungen aus meinem Leben* (Schaffhausen, 1863–4)

J. Cox: *Musical Recollections of the Last Half-century* (London, 1872)

J. R. Planché: *Recollections and Reflections*, i (London, 1872, rev. 2/1901)

C. Moscheles, ed.: *Aus Moscheles Leben* (Leipzig, 1872–3; Eng. trans., 1873)

F. Kemble: *Record of a Girlhood* (London, 1878, 2/1879)

E. Genast: *Erinnerungen aus Weimars klassischer und nachklassischer Zeit* (Stuttgart, 2/1887, 5/1905)

P. S. Allen and J. T. Hatfield, eds.: *Wilhelm Müller: Diary and Letters* (Chicago, 1903) [Ger. orig.]

E. Michotte: *Souvenirs personnels: la visite de R. Wagner à Rossini* (Paris, 1906; Eng. trans., 1968, ed. H. Weinstock)

H. and C. Cox, eds.: *Leaves from the Journals of Sir George Smart* (London, 1907)

C. Costenoble: *Tagebücher von seiner Jugend* (Berlin, 1912)

P. Frederick, ed.: *F. W. Gubitz: Bilder aus Romantik und Biedermeier* (Berlin, 1922)

F. Baser: 'Mannheimer-Heidelberger Freundschaft zur Zeit der Romantik: Weber und Freiherr Alexander von Dusch', *Die Musik*, xxvi (1933–4), 277

L. Schmidt: 'Zeitgenössische Nachrichten über Carl Maria von Weber', *Die Musik*, xxxiii (1940–41), 653

M. Gregor-Dellin, ed.: *Richard Wagner: Mein Leben* (Munich, 1963)

80

Bibliography

LIFE AND WORKS

J. Benedict: *Weber* (London, 1881, 5/1899)

P. Spitta: 'Weber', *Grove 1–4*

H. Gehrmann: *Carl Maria von Weber* (Berlin, 1899)

B. Schuster, ed.: *Die Musik*, v (1905–6), no.17 [Weber issue]

J. Kapp: *Weber* (Stuttgart and Berlin, 1922, 15/1944)

A. Coeuroy: *Weber* (Paris, 1925, 2/1953)

E. Kroll: *Carl Maria von Weber* (Potsdam, 1934)

H. Dünnebeil: *Carl Maria von Weber: Leben und Wirken dargestellt in chronologischer Tafel* (Berlin, 1953)

H. Schnoor: *Weber: Gestalt und Schöpfung* (Dresden, 1953)

K. Laux: *Carl Maria von Weber* (Leipzig, 1966)

J. Warrack: *Carl Maria von Weber* (London, 1968, 2/1976)

BIOGRAPHICAL STUDIES

Obituary, *Quarterly Musical Magazine and Review*, viii (1826), 121

M. M. von Weber: *Carl Maria von Weber: ein Lebensbild* (Leipzig, 1864–6, abridged 2/1912 ed. R. Pechel; Eng. trans., abridged, 1865/R1968)

F. W. Jähns: *Carl Maria von Weber: eine Lebensskizze nach authentischen Quellen* (Leipzig, 1873)

A. Jullien: 'Weber à Paris en 1826', *Revue et gazette musicale*, xxxv (1877), 17–57; pubd separately (Paris, 1877); repr. in *Paris dilettante au commencement du siècle* (Paris, 1884)

K. Knebel: 'Weber in Freiberg 1800–1', *Mitteilungen vom Freiberger Altertumsverein*, xxxvii (1900), 72

H. Pohl: 'Drei Aktenstücke über Webers Gefangensnehme in Stuttgart, 1810', *AMz*, xxxvii (1910), 1207

J. Kapp: 'Carl Maria von Weber in Prag', *Die Musik*, xv (1922–3), 117

F. Walter: 'Weber in Mannheim und Heidelberg 1810 und sein Freundeskreis', *Mannheimer Geschichtsblätter*, xxv (1924), cols.18–73

K. Esselborn: 'Weber und Darmstadt', *Darmstädter Blätter für Theater und Kunst*, iv (1925–6), 209

E. Abrahamsen: 'Carl Maria von Weber in Copenhagen', *The Chesterian*, vii (1926), 227

F. Hefele: *Die Vorfahren Karl Maria von Webers* (Karlsruhe, 1926)

M. Herre: 'Weber und Augsburg', *Zeitschrift des historischen Vereins für Schwaben und Neuburg*, xlvii (1927), 217

O. Huffschmid: 'Karl Maria von Weber in Heidelberg', *Festspielbuch der Heidelberger Festspiele* (1928), 124

E. Schenk: 'Über Weber's Salzburger Aufenthalt', *ZMw*, xi (1928–9), 59

H. Fritsche: 'Weber in Carlsruhe, Oberschlesien', *Schlesische Monatsheft*, xii (1936), 417

81

——: 'Webers Breslauer Zeit', *Schlesische Monatsheft*, xii (1936), 410

O. Kroll: 'Weber und Baermann', *NZM*, Jg.103 (1936), 1439

Z. Němec: *Weberová pražská léta* [Weber's Prague years] (Prague, 1944) [with facs. and Cz. transcr. of Prague notebook]

STAGE WORKS: CRITICAL AND HISTORICAL STUDIES

C. Brühl: *Neueste Kostüme auf beiden königlichen Theatern in Berlin* (Berlin, 1822) [incl. *Freischütz* costumes]

'Oberon, or The Elf King's Oath', *Quarterly Musical Magazine and Review*, viii (1826), 84

H. von Chezy: 'Carl Maria von Webers Euryanthe: ein Beitrag zur Geschichte der deutschen Oper', *NZM*, xiii (1840), 1, 9

F. Kind: *Freischütz-Buch* (Leipzig, 1843) [incl. first-hand account of opera's composition]

H. Berlioz: 'Le Freyschütz de Weber', *Voyage musicale en Allemagne et en Italie*, i (Paris, 1844), 369

R. Schumann: *Gesammelte Schriften über Musik und Musiker* (Leipzig, 1854, 4/1914)

R. Wagner: 'Der Freischütz in Paris', *Gesammelte Schriften und Dichtungen*, i (Leipzig, 1871); *Prose Works*, ed. W. Ashton Ellis, vii (London, 1898)

G. Servières: *Freischütz* (Paris, 1913) [Fr. trans. of lib; discussion of Fr. versions]

E. Hasselberg, ed.: *Der Freischütz: Friedrich Kinds Operndichtung und ihre Quellen* (Berlin, 1921)

J. Kapp: 'Die Uraufführung des Freischütz', *Blätter der Staatsoper*, i/8 (Berlin, 1921), 9

H. Abert: 'Carl Maria von Weber und sein Freischütz', *JbMP 1926*, 9; repr. in *Gesammelte Schriften und Vorträge*, ed. H. Blume (Halle, 1929, 2/1968)

S. Goslich: *Beiträge zur Geschichte der deutschen romantischen Oper* (Leipzig, 1937)

T. Cornelissen: *Carl Maria von Webers Freischütz als Beispiel einer Opernbehandlung* (Berlin, 1940)

H. Schnoor: *Weber auf dem Welttheater: ein Freischützbuch* (Dresden, 1942, 4/1963)

R. Engländer: 'The Struggle between German and Italian Opera at the Time of Weber', *MQ*, xxxi (1945), 479

P. Kirby: 'Weber's Operas in London, 1824–6', *MQ*, xxxii (1946), 333

W. Kron: *Die angeblichen Freischütz-Kritiken E. T. A. Hoffmanns* (Munich, 1957)

G. Mayerhofer: *Abermals vom Freischützen: der Münchener Freischütze von 1812* (Regensburg, 1959)

A. A. Abert: 'Webers Euryanthe und Spohrs Jessonda als grosse Opern', *Festschrift für Walter Wiora* (Kassel, 1967), 435

Bibliography

G. Jones: *Backgrounds and Themes of the Operas of Carl Maria von Weber* (diss., Cornell U., 1972)

S. Goslich: *Die deutsche romantische Oper* (Tutzing, 1975)

G. Jones: 'Weber's "Secondary Worlds"': the Later Operas of Carl Maria von Weber', *IRASM*, vii (1976), 219

K. Laux: 'In Erinnerung gebracht', *Musikbühne 76*, ed. H. Seeger (Berlin, 1976), 89 [on *Die drei Pintos*]

J. Warrack: '*Oberon* und der englische Geschmack', *Musikbühne 76*, ed. H. Seeger (Berlin, 1976), 15 [with facs. of complete *Oberon* text in Planché's first edn.]

A. Csampai and D. Holland, eds.: *Carl Maria von Weber: Der Freischütz* (Reinbek bei Hamburg, 1981)

OTHER WORKS: CRITICAL AND HISTORICAL STUDIES

G. Kaiser: *Beiträge zu einer Charakteristik Carl Maria von Webers als Musikschriftsteller* (Berlin, 1910)

H. Allekotte: *Carl Maria von Webers Messen* (Bonn, 1913)

W. Georgii: *Carl Maria von Weber als Klavierkomponist* (Leipzig, 1914)

M. Degen: *Die Lieder von Carl Maria von Weber* (Freiburg, 1923)

R. Tenschert: 'Die Sinfonien Webers', *Neue Musik-Zeitung*, xlviii (1927), 481

G. Abraham: 'Weber as Novelist and Critic', *MQ*, xx (1934), 27; repr. in *Slavonic and Romantic Music* (London, 1968)

P. Listl: *Weber als Ouvertürenkomponist* (Würzburg, 1936)

MISCELLANEOUS STUDIES

A. Krüger: *Pseudoromantik: Friedrich Kind und der Dresdner Liederkreis* (Leipzig, 1904)

E. Kroll: 'E. T. A. Hoffmann und Weber', *Neue Musik-Zeitung*, xlii (1921), 336

R. Godet: 'Weber and Debussy', *The Chesterian*, vii (1926), 220

H. Pfitzner: *Was ist uns Weber?* (Augsburg and Cologne, 1926)

C. Wagner: 'Weber und die Lithographie', *Buch und Schrift: Jb des deutschen Vereins für Buchwesen und Schrifttum*, v (1931), 9

E. Kroll: 'Beethoven und Weber', *NBJb*, vi (1935), 124

E. Brandt: 'Franz Anton von Weber als Leiter der Eutiner Hofkapelle', *NZM*, Jg.103 (1936), 1439

W. Muche: 'Weber und Marschner', *Signale für die musikalische Welt*, xciv (1936), 746

W. Thomas: 'Carl Maria von Weber als Musikpolitiker', *NZM*, Jg. 103 (1936), 1443

P. Raabe: *Wege zu Weber* (Regensburg, 1942)

P. Kirby: 'Washington Irving, Barham Livius and Weber', *ML*, xxxi (1950), 133

83

G. Hausswald, ed.: *Carl Maria von Weber: ein Gedenkschrift* (Dresden, 1951)

W. Becker: *Die deutsche Oper in Dresden unter der Leitung von Carl Maria von Weber, 1817–26* (Berlin, 1962)

D. Reynolds, ed.: *Weber in London, 1826* (London, 1976)

HECTOR BERLIOZ

Hugh Macdonald

CHAPTER ONE

Life

I 1803–21

Louis-Hector Berlioz, born on 11 December 1803, was the eldest child of Louis-Joseph Berlioz (1776–1848), a doctor of some repute and a prominent, well-to-do citizen of La Côte-St-André, 48 km north-west of Grenoble in the département of Isère. The family had belonged to the region for many generations, and the countryside, especially the grandeur of the Isère plain against its distant background of the Alps, cast a lasting spell on the young composer. His father was a man of liberal outlook and broad intellectual range, an inspiring mentor for his son, and though the dispute over Hector's career and marriage damaged their relationship for some years, there was a profound bond between them. His mother, Marie-Antoinette (née Marmion), was a Catholic of sharper temper and narrower outlook. Five more children were born, of whom two, Nanci and Adèle, lived to maturity, and enjoyed Berlioz's permanent affection.

At the age of ten Berlioz briefly attended an infant seminary at La Côte, but thereafter his education was entirely in his father's hands. He took most keenly to French and Latin literature and to geography, especially travel books, which implanted in him a longing for distant, sometimes exotic shores that his later travels around Europe scarcely satisfied. Of Latin authors his favourite was Virgil, and in his *Mémoires* (the major source for knowledge and understanding of his life) he recounted how his father's reading of the episode of Dido and Aeneas reduced him to tears. His father also gave him rudimentary instruction on the flageolet, and he later learnt the flute with a local teacher, Imbert, and the guitar with another, Dorant. There is no doubt that his ability on the flute and guitar quickly became more than adequate, and it satisfied not only social demand but the deep-rooted sensitivity to music of which Berlioz had first become aware as a small boy when attending Mass. A tune from Dalayrac's *Nina*, pressed to the service of religion, first evoked a sense of wonderment mingling with an ardent but short-lived religious sense. He never studied the piano and never learnt to play more than a few chords.

Berlioz also linked his first steps in music, learning the flageolet, with his boyhood passion for Estelle Duboeuf, when he was 12 and she 18. He called her his *stella montis*, associating her with the mountains behind Meylan where she lived and with Florian's *Estelle et Némorin* which he had already 'read and reread a hundred times'. He was teased for his admiration from afar, but it proved to be deeper than anyone suspected. He found a copy of Rameau's *Traité de l'harmonie* at home and also procured Catel's *Traité d'harmonie*. These provided the basis for a knowledge of harmony, learnt entirely without reference to a keyboard, with which he began to compose more ambitiously, probably at the age of 13 or 14. He wrote a potpourri on Italian melodies, and then two quintets for flute and strings, all now lost but for a melody from one of the quintets that became the second subject of the overture to *Les francs-*

juges. Similarly, a setting of one of Florian's poems, *Je vais donc quitter pour jamais*, made at this time survives as the first theme in the opening section of the *Symphonie fantastique*. 'It seemed to me exactly right for expressing the overpowering sadness of a young heart caught in the toils of a hopeless love.' He made copies of popular *romances* by Dalayrac, Boieldieu, Berton and others, sometimes with his own guitar accompaniments, and his own *romances* were in the same mould. In 1819, when he was 15, he wrote to two (and probably more) Paris music publishers offering a sextet and some songs with piano accompaniment, and it seems likely that *Le dépit de la bergère*, a *romance* and his first published work, was accepted in this way. Though unambitious in style it distantly prefigures the *Sicilienne* in *Béatrice et Bénédict*, composed over 40 years later.

When he reached 17 a decision had to be made about his career, and though an irresistible instinct drew Berlioz to music, his father's wish that he should follow him into the medical profession prevailed, and he was sent to Paris to the École de Médecine, having obtained his bachelor's degree in Grenoble in March 1821. At this stage his horizons were still narrow; his knowledge of the world was more literary than real, and his profoundest impressions were probably the child's absorption of his natural surroundings and of the echoes of the Napoleonic convulsion. Much of his experience was vicarious, for he found in Bernardin de St Pierre and Chateaubriand an outlet for his still dormant capacity for intense feeling. In music only the slightest works by minor composers were known to him and he had never seen a full score; Pleyel's quartets were the most sophisticated music he had heard. In physique he was of middle height, with a mass of fiery, tawny hair; his eyes were blue and deep set, and a distinctive aquiline nose surmounted wide, thin lips.

II 1821–30

Even before Berlioz's departure from La Côte his aversion to medicine was plain.

Become a doctor! Study anatomy! Dissect! Take part in horrible operations – instead of giving myself body and soul to music, sublime art whose grandeur I was beginning to perceive! Forsake the highest heaven for the wretchedest regions of earth, the immortal spirits of poetry and love and their divinely inspired strains for dirty hospital orderlies, dreadful dissecting-room attendants, hideous corpses, the screams of patients, the groans and rattling breath of the dying! No, no! It seemed to me the reversal of the whole natural order of my existence. It was monstrous. It could not happen. Yet it did.

With his cousin Alphonse Robert, with whom he shared lodgings, he attended medical school and pursued his studies for two years, at least until his baccalauréat de sciences physiques which he took in January 1824.

But medicine was fighting a losing battle against the overpowering strength of Berlioz's musical impulse now inflamed a hundred times more strongly by the musical experience and opportunity offered by the capital, which was to remain his home for the rest of his life. Within a month of his arrival he began to attend performances at the Opéra. Gluck, whose *Iphigénie en Tauride* was one of the first operas he heard, made a deep and lasting impression and remained the composer he admired most wholeheartedly of all. He also heard operas by Salieri, Sacchini, Méhul, Spontini and Boieldieu, a repertory that supplied a stylistic basis for his own initial attempts at large-scale composition. There survive copies in his own hand of extracts from Gluck's operas made in 1822 in the Conservatoire library, which he frequented as often as his studies allowed, but he soon felt the need to supplement his musical technique; at the end of 1822 he gained an introduction to Le Sueur through a pupil, Gerono, and was admitted to his class. By this time he had attempted for the first time a work for full orchestra, the cantata *Le cheval arabe* on a text by Millevoye (now lost). Six *romances* for one or two voices with piano had appeared separately in print since his arrival in Paris, but one effect of Le Sueur's tutelage was that Berlioz published no further music for about six years, concentrating instead on larger works, with orchestral accompaniments. In 1823 he composed an opera on Florian's *Estelle et Némorin* referring to childhood memories and doubtless childhood melodies too. This, like the two works that followed – a scene for bass from Saurin's *Beverley* and the Latin oratorio *Le passage de la mer rouge* – was later burnt, on Berlioz's confession. The first important work to have survived even in part is the mass composed for the church of St Roch in 1824. A first rehearsal under Valentino on 27 December 1824 was a fiasco, but a successful performance the following July, under the same conductor, restored Berlioz's confidence in himself and strengthened his resolve to be, in Le Sueur's words, 'no doctor or apothecary but a great composer'.

Since his abandonment of medicine he had had to face the entrenched opposition of his parents and their curtailment of his funds. Family disputes persisted for years and the visits he paid to La Côte only deepened the estrangement. With his father's allowance reduced and intermittently refused, Berlioz was forced to borrow from his friends, and he was to suffer severe hardship for at least five years. He depended on whatever sources were at hand – a few pupils, a short period as a chorus singer at the Théâtre des Nouveautés, and occasional articles for the press, the beginnings of what was later to be his principal source of income. His closest friend at this period was a law student with literary inclinations, Humbert Ferrand, who supplied the text for *La révolution grecque*, set to music in 1825, and also for an opera *Les francs-juges* (1826) of which six fragments and the overture remain. It owed much to his teacher Le Sueur but also reflected the dark colours and sinister tones of Weber's *Der Freischütz* (introduced to Paris as *Robin des bois* the previous year). Although *Les francs-juges* was completed, none of Berlioz's efforts to secure a performance succeeded, and after at least two attempts to rewrite it, he discarded or re-used some of it in later works, notably the *Marche des gardes*, incorporated in the *Symphonie fantastique* as the 'Marche au supplice' in 1830.

In 1826 Berlioz entered the Conservatoire: here he was in Le Sueur's class for composition, which he had been attending for some time, and Reicha's class for counterpoint and fugue. *Les francs-juges* and the overture to *Waverley*, which followed it, revealed a growing individuality and a marked confidence in his own powers, especially in the handling of instruments. Membership of the Conservatoire entitled him to enter for the annual Prix de Rome offered by the institute for musical composition, and he entered *en loge* in 1827 for the first of four times. *La mort d'Orphée*, the cantata set in 1827, was declared unplayable by the judges (though Berlioz had it played in rehearsal in 1828 with some satisfaction). For the 1828 competition he composed *Herminie*, which contains the melody later used as the *idée fixe* in the *Symphonie fantastique*, and won second prize. In 1829 he wrote the most individual and dramatic of these cantatas, *La mort de Cléopâtre*, and no prize was awarded, probably to avoid bestowing official approval on a composer who 'betrayed such dangerous tendencies'. At the fourth attempt, in 1830, Berlioz was finally successful, although only a fragment of the cantata, *La mort de Sardanapale*, survives. His tactic had been to restrain his more individual mode of expression in order to provide a conventionally acceptable style.

Meanwhile the emotional and artistic elements of his being had been set alight by a series of thunderstrokes. The capacity for absorbing powerful external impressions and transmuting them into high artistic form placed him in the avant garde of the generation of 1830, and implanted in the soil of his imagination the seed of great works, many of them to remain beneath the surface of realization for many years. The first was the simultaneous impact of Shakespeare and the actress Harriet Smithson on 11 September 1827. On that day Berlioz attended *Hamlet* presented by an English company at the Odéon theatre, with Charles Kemble playing Hamlet and Miss Smithson playing Ophelia. 'The impression made on my heart and mind by her extraordinary talent, nay her dramatic genius, was equalled only by the havoc wrought in me by the poet she so nobly interpreted.' Though the performance (of the Garrick version) was in English, of which Berlioz knew virtually nothing at that time, he grasped the grandeur and sublimity of Shakespeare's language and the richness of its dramatic design, and he joined the ranks of those under

8. Hector Berlioz: portrait (1830) by Emile Signol

Hugo's leadership who extolled Shakespeare as a challenge to French Classicism and a model for the new Romantic theatre. For Berlioz Shakespeare represented the pinnacle of poetic utterance; his veracity of dramatic expression and freedom from formal constraints picked up direct resonances in Berlioz's spirit. Shakespeare's plays were to supply the basis of three major works, *Romeo et Juliette*, *Béatrice et Bénédict* and the *Roi Lear* overture. In addition, there were at least three pieces inspired by *Hamlet*, a fantasy on *The Tempest*, and some direct borrowings in *Les troyens*. More profoundly Shakespeare provided a framework for the structure of both *Roméo et Juliette* and *Les troyens* and was a source, in the form of dramatic truth, of Berlioz's fundamental notion of expressive truth. Berlioz was to read and quote Shakespeare avidly for the rest of his life, putting him alongside Virgil in his literary pantheon.

This seminal discovery worked itself out more profoundly and more slowly than that of Miss Smithson, whom he referred to as his Ophelia, or Juliet, or Desdemona. His emotional derangement was immediate and violent. For the next two years he pursued her remorselessly, waiting for her return to Paris, vainly seeking a means to approach her. When in 1830 it seemed that his love for her had turned sour, the accumulation of emotional tension broke out in the *Symphonie fantastique*, which describes and transmutes into artistic form the artist's passions, dreams and frustrations. For Berlioz there was no clear distinction between the real Harriet Smithson and the idealized embodiment of Shakespeare's heroines, so that when, later, he was to secure an introduction to her and ultimately marry her, a relationship that had begun on an ideal level could only spoil in the glare of everyday reality, and the wholly Romantic conjunction of the artist with the ideal woman came to a bitter end.

Two further discoveries at this time rank as of supreme importance: in March 1828 Berlioz heard Beethoven's Third and Fifth Symphonies, played by Habeneck and the Société des Concerts at the Conservatoire. 'Beethoven opened before me a new world of music, as Shakespeare had revealed a new universe of poetry.' For the first time his horizons widened from the exclusively vocal genres of opera, cantata and *romance* to the expressive potential of pure instrumental music. That Berlioz wrote symphonies at all is entirely due to his obeisance to Beethoven, and the *Symphonie fantastique* can be seen as a deliberate and conscious attempt to work out dramatic and poetic ideas in the framework of a Beethoven symphony. More important, Berlioz discovered that instrumental music has an expressive and articulative force far more penetrating than vocal setting, a discovery shown palpably in the 'Scène d'amour' of *Roméo et Juliette*, in the *Hamlet* funeral march, and at certain points in *Les troyens*. Just as Berlioz hardly set any of Shakespeare's poetry to music, similarly Berlioz rarely adopted the precise tone and timbre of Beethoven. He absorbed this impact at a deep level, seeing Beethoven as a supreme dramatist in music, more poet than craftsman.

Goethe's *Faust* reached Berlioz through Gérard de Nerval's translation, published in December 1827, and again its impact was profound and immediate. The Faustian conception of man struck numerous echoes in Berlioz's breast. In a letter of 1828 he described Shakespeare and Goethe as 'the silent confidants of my suffering; they hold the key to my life'. He went on to say that he had just set the ballad of the King of Thule to music, the first of what were to be eight scenes, settings of the verse portions of Nerval's translation. The *Huit scènes de Faust* were published at Berlioz's expense in the following year, an op.1 of exceptional originality and invention. Each scene bears a quotation from Shakespeare, and each has its appropriate musical setting, varying from the *Concert de sylphes* for six solo voices and orchestra to the *Sérénade* in which Mephistopheles is accompanied by a guitar. But despite its remarkable character Berlioz found the work 'crude and badly written'. He collected all the copies he could and destroyed them. Dimly he may have realized that the music would eventually find its due place in the larger scheme of *La damnation de Faust*, completed in 1846.

Literary influences of a less overwhelming kind were numerous, chief among them being the works of Moore, Scott and Byron. All three inspired compositions. He submerged himself, too, in Chateaubriand, Hoffmann, Fenimore Cooper, and the work of his own compatriots and contemporaries, Hugo, de Vigny, de Musset and Nerval. Later he was to absorb and admire Balzac, Flaubert, and Gautier, whose poems supplied the text of *Les nuits d'été*.

The ferment of Berlioz's mind in the late 1820s was astonishing. Instead of wilting under a constant onslaught on his sensitivities, he broke out in gusts of creative energy. The *Waverley* overture, the *Huit scènes de Faust*, the nine settings of Thomas Moore (the *Irlande* collection), composed in 1829, and above all the *Symphonie fantastique*, composed in early 1830, are

97

testimony to this. He was active too as a proponent of his own music. The mass had been played at St Roch in 1825 and 1827, and on 26 May 1828 Berlioz gave his first orchestral concert in Paris. His intention was to bring himself to the attention of the public, especially Harriet Smithson, and he succeeded in his aim in that the press, particularly the influential Fétis, was favourable. Much of his time in the next 15 years was devoted to planning, organizing and, after 1835, conducting his own concerts in Paris, a task that made heavy demands on his energy and usually his purse, but which provided the sole outlet for his orchestral works.

His eventual success in winning the Prix de Rome in 1830 was important to him as a means of convincing his parents that his musical bent was serious, as well as a source of income for the next few years. The prize required residence in Italy, but before he left he had important concerts to give in Paris. At the prize-giving ceremony at the institute on 30 October his cantata *La mort de Sardanapale* (with an additional conflagration scene written after the prize had been awarded) misfired completely. His fantasy (or 'overture') on *The Tempest* for chorus and orchestra was heard for the first time on 7 November at the Opéra and on 5 December the *Symphonie fantastique* received its first performance in a concert of Berlioz's works conducted by Habeneck. Liszt, who was present, made Berlioz's acquaintance on that occasion (see fig.9, p.100).

Berlioz's reputation as a composer of startling originality was by now confirmed and his progress in the musical world of Paris was not to be furthered by enforced removal to Italy. He made several requests to be exempted from going, giving as his reasons his need to pursue his career in Paris and the state of his health, which had certainly not been good. A more pressing reason, in Berlioz's mind, was his attachment to Camille Moke, a 19-year-old pianist of exceptional gifts whom he had met earlier in the year at a school where she taught the piano and he the guitar. She replaced the unresponsive Miss Smithson in his affections and their ardent affair led even to betrothal on the eve of his departure for Italy.

III   1831–42

Berlioz spent a month at La Côte-St-André, where his parents were at last delighted with the success he had achieved. At the back of his mind he already had a large-scale composition that was to haunt him for a number of years, while his immediate thoughts were entirely with Camille, already, according to Ferdinand Hiller (her previous attachment), cooling in her affections. His journey to Italy and the 15 months he spent there were crucially formative. His mind was constantly alive to the impressions, both inspiring and disappointing, of the country and the people, their customs and way of life. He was supposed to draw inspiration from the relics of classical antiquity. These certainly intrigued him, especially where they touched upon Virgil, but his musical output was relatively small and haphazard, and his official submissions from Rome were not especially remarkable. Italy was nonetheless to work upon his music in more gradual fashion, with far-reaching influence on his style. Henceforth there was a new colour and glow in his music, both sensuous and vivacious. These derive not from Italian art, which touched him little, or Italian music, which he despised, but from the scenery and the sun, and from his acute sense of locale. *Harold en Italie*, *Benvenuto Cellini* and *Roméo et Juliette* are the most obvious expressions of his response to Italy: both *Les troyens* and *Béatrice et Bénédict* reflect the warmth and stillness of the Mediterranean, as well as its vivacity and force.

Berlioz's descriptions of Italy in the *Mémoires* and the abundant accounts of his travels in letters to his friends and family are wonderfully evocative; he seems here to have discovered his gifts as a writer. In Italy he came face to face with experiences he had previously only read about or idealized. Byronism, so fashionable at that time, became reality as he encountered brigands, corsairs, revolutionaries, *lazzaroni* and *pifferari*, and as he sampled the harshness of a storm at sea or the Carnival in Rome or sleeping in the open air in the mountains. He met sailors, peasants, sculptors and travellers, but, with the notable exception of Mendelssohn, few musicians. The Villa Medici at Rome housed the institute prizewinners under the tutelage of Horace Vernet, but Berlioz greatly disliked the city: 'Rome is the most stupid and prosaic city I know: it is no place for anyone with head or heart'. Florence, on the other hand, he adored: 'Everything about it delights me, its name, its climate, its river, its palaces, its air, the style and elegance of its inhabitants, its surroundings, everything, I love it, love it'. At Rome he composed little, mostly because of the stifling atmosphere of the Villa, but on his travels he achieved much more.

Three weeks after his arrival in Rome Berlioz set off back to France, jeopardizing his pension, in order to discover why he had heard nothing from Camille. At Florence, where he suffered a serious attack of quinsy,

9. Autograph MS from Chapter 31 of Berlioz's 'Mémoires'; in this passage the composer describes his first meeting with Liszt and reactions to the première of the 'Symphonie fantastique' on 5 December 1830 he learned the truth: that she had abandoned him for a new and more prosperous suitor, Camille Pleyel, the piano manufacturer. In a torrent of rage and wounded pride, Berlioz determined to return to Paris to kill the two Camilles, her mother and finally himself. Though he reached Nice his resolve wavered and his better sense persuaded him to give his passions time to cool. Vernet was prepared to pardon him; Berlioz was prepared to spare his victims. The experience was traumatic, with emotional recovery very closely related to the recovery of his health. He felt that he had 'survived', and that he could live again to compose the music still dormant in his mind. Here was born *Le retour à la vie*, a half-literary, half-musical work that folded a variety of experiences together under the title 'mélologue' taken from Thomas Moore. Much of the text reflected thoughts and ideas found in his letters of the time, while the music was drawn mostly from works written earlier in Paris. Although the work was always designed as a sequel to the *Symphonie fantastique* and referred directly to Harriet Smithson, it was a different unrequited love that originated it. It was not renamed *Lélio* until its revival in 1855.

Resting for three weeks in Nice – the three happiest weeks of his life, Berlioz said – he gave priority to another pressing inspiration, an overture on Shakespeare's *King Lear*, which he had read in Florence, and started another on Scott's *Rob Roy*. On the return to Rome he worked further on *Le retour à la vie* and revised the *Symphonie fantastique*. He moved out of Rome as often as possible, especially to the Abruzzi mountains, Tivoli and Subiaco, where he finished *Rob Roy*. Antoine Etex, the sculptor, recalled how he and Berlioz had gone for long walks together, singing *Guillaume Tell*, bathing, searching for brigands and playing practical jokes. In September, Berlioz went to Naples, visited Pompeii and the island of Nisida, and then returned to Rome on foot.

The only musical product of the rest of his stay in Italy was the song *La captive* (in its strophic form), written in Subiaco in February 1832. Impatient to get back to Paris and to have his new works performed there, he secured six months' dispensation and left in May 1832. He was later exempted from the required residence in Germany. After some months at La Côte-St-André, he reached Paris in November and immediately organized a concert of his own music, including the revised *Symphonie fantastique* with its sequel *Le retour à la vie*. Thinly veiled references to Fétis's 'corrections' of Beethoven symphonies earned Berlioz a bitter notice in the *Revue musicale* and the animosity of an influential critic, but he was more concerned by the fact that Harriet Smithson was in Paris at the time. Her performances were far from the fashionable success they had been five years earlier, but an intermediary secured her attendance at the concert and subsequently an introduction. Despite their respective accumulated debts and difficulties and objections from both families, especially his own, Berlioz soon proposed marriage. After a bizarre and stormy courtship, they were married on 3 October 1833. It was perhaps characteristic of Berlioz to take his idealized love for his Ophelia to the point of marriage and perhaps, too, no surprise that the marriage was happy for scarcely more than six years. A son, Louis, was born to them in August 1834, and the picture of the young ménage living at the top of

Montmartre, where their friends went to visit them, is a touching one. But with a language barrier between them and the strains of temperament and material deprivations always acute, it is hardly surprising that by 1842 or earlier they had drifted apart. Harriet's last years are a distressing tale of misery and decline, and she died in 1854. Berlioz supported her to the end and retained a warmth of affection for what she had meant to him and for her inspiring qualities as an artist.

Berlioz's career in the 1830s is, despite its astonishing achievements, essentially a tragic one. Conscious of his own genius and of the springs of invention within him, he failed to win the recognition that alone assured him even the barest means of existence. As a composer he earned virtually nothing. The general view of his music was that it was eccentric and 'incorrect'. His admirers were passionate but few, and no worthy official post, such as a teaching appointment at the Conservatoire, came his way; he became merely its assistant librarian. He secured two government commissions (for the *Grande messe des morts* in 1837 and the *Grande symphonie funèbre et triomphale* in 1840), but neither of these was particularly profitable or helpful to his artistic standing. He was compelled to earn a living in a profession at which he excelled but which he abhorred – as a critic. He wrote for *L'Europe littéraire* in 1833, *Le rénovateur* from 1833 to 1835, and principally from 1834 for the *Gazette musicale* (later to become the *Revue et gazette musicale*), and the *Journal des débats*, an influential newspaper whose proprietors, the Bertins, were his staunch supporters. He was soon to be better known to Parisians as a critic than as a composer.

Journalism took him away from composing and from

its essential adjunct, performance. Gladly would Berlioz have devoted more of his time to giving concerts, even though the financial burdens were always severe. The record of his concerts in Paris is as follows: in 1832 he gave two, in 1833 five, in 1834 four, in 1835 five, in 1836 two, in 1837 the official performance of the *Grande messe des morts*, in 1838 two, in 1839 three, in 1840 five, in 1841 one and in 1842 four. The programme was normally made up of his own music interspersed with vocal and instrumental solos and occasionally works by Beethoven, Weber, Spontini and others. Liszt, Chopin, Hallé and other members of the richly cosmopolitan circle of musicians who then inhabited Paris took part. After 1835, when Girard bungled a performance of *Harold en Italie*, Berlioz resolved to conduct his own works himself. This led, in turn, to a further career as one of the first specialist orchestral conductors, in wide demand outside France for his skill and interpretative insight.

Discouragement could not stem the flow of major compositions. *Harold en Italie* was composed in the summer of 1834 in response to a request from Paganini for a work in which he might display a fine Stradivari viola. Berlioz used the opportunity to devise an unusual symphony with concerto elements in which echoes of his Italian journey are presented in the cloak of Byron's *Childe Harold*. As in the *Symphonie fantastique*, a recurrent theme again serves to unify the four movements, but the modest role of the viola solo deterred Paganini from ever playing it.

If the image of Beethoven was still vivid in Berlioz's mind at this time, his primary concern, for professional as well as artistic reasons, was to win success at the

Opéra. Only thus was real recognition to be sought; only thus, too, could Berlioz prove himself in the noble line of Gluck and Spontini. *Les francs-juges* had already been revised once before he left for Italy. After his return he made a further attempt, with Thomas Gounet's help, to refashion it into a single act, but it still aroused no interest. After abandoning it, he considered a comic opera on Shakespeare's *Much Ado About Nothing* (this eventually materialized as *Béatrice et Bénédict* in 1862) and briefly contemplated *Hamlet* before persuading Léon de Wailly and Auguste Barbier, with Alfred de Vigny's assistance, to make a libretto out of Benvenuto Cellini's *Vita* (*Memoirs*), a book whose abundant incident and ardent idealization of the artist–hero appealed strongly to him. It provided, too, the irresistible local colour of Renaissance Italy. First written as an *opéra comique* with dialogue, the libretto was refused in 1834, but by elevating the tone and extending the music Berlioz was able to offer it to the Opéra. It was accepted in 1836 and performed in 1838. At that time the music of Meyerbeer and Halévy held such sway at the Opéra that few members of the company were able or prepared to consider Berlioz's bewilderingly original and inventive music with real seriousness. At all events, the three performances of 1838 were a clear failure, and the management had little interest in the few fragmentary revivals the following year. Berlioz described the experience with bitterness as being 'stretched on the rack', for it not only humiliated him as an artist, it also closed the door of the Opéra to him, except as the arranger of other men's works, for the rest of his life.

Berlioz was preoccupied at the same time with a half-Revolutionary, half-Napoleonic conception on the grandest scale, which took various forms. Remnants of the 1824 mass, a military symphony sketched out on the journey back from Italy, and a preoccupation with the Last Judgment all contributed to plans for a huge work in seven movements commemorating France's national heroes, of which two movements were completed in 1835. These do not survive, although they were probably included in the Requiem commissioned by the minister of the interior and performed in the Invalides on 5 December 1837, and also perhaps in the *Grande symphonie funèbre et triomphale*, another government commission, performed during the tenth anniversary of the 1830 Revolution on 28 July 1840. Both works exploit Berlioz's interest in grandiose spatial effects and in the appropriate matching of instrumental forces to the occasion and the place for which a piece was intended. The *Symphonie funèbre* was originally written for large military band and performed out of doors. Berlioz later added parts for strings and for chorus. Traces of his Napoleonic leanings may be seen in his setting for solo bass, chorus and orchestra of Béranger's *Le cinq mai*, first performed in 1835.

In contrast to these, many compositions of the 1830s were delicate and intimate. He continued to write songs, of which some were orchestrated, such as *La captive* and *Le jeune pâtre breton*. *Sara la baigneuse*, an exceptionally refined setting of a Hugo poem, was first heard in 1834. *Les nuits d'été*, six settings of Gautier poems with piano accompaniment, appeared in 1841: all six were later orchestrated.

Paganini's unexpected gift of 20,000 francs in December 1838, a token of his admiration for *Harold en Italie*, made possible the composition of *Roméo et*

107

10. Hector Berlioz: pencil portrait (c1832) attributed to Jean Auguste Dominique Ingres

Juliette, and consoled Berlioz for the failure of *Benvenuto Cellini*. 'My one idea was to put it to a musical purpose. I would give up everything else and write a really important work, something splendid on a grand and original plan, full of passion and imagination, worthy to be dedicated to the glorious artist to whom I owed so much.' Berlioz wrote movingly of the ardent months of composition and he came to regard the 'Scène d'amour' as one of his finest things, but the critics accused him of failing to understand Shakespeare. For Wagner at least, who was present at the first performance, it was a 'revelation'.

IV **1842–8**
About 1841 Berlioz reached a turning-point in his career. In that year the only music of his that was publicly performed in Paris was the set of recitatives composed for Weber's *Der Freischütz* in order to make it acceptable to the Opéra's ban on spoken dialogue. At the same time reports of performances abroad were increasingly common. The Requiem, for example, was heard in St Petersburg, while smaller works, such as the overtures, especially *Les francs-juges*, were becoming more frequently heard in England and Germany. He still withheld publication of the symphonies to prevent performances outside his control, so that it was growing urgent to go abroad in person, and to reinforce a developing international reputation. At the same time the frustrations of Paris made themselves more keenly felt, with the brighter enthusiasms of 1830 already receding and bourgeois tastes daily more evident, especially in the theatre. His marriage was perhaps already strained. For the first time his musical creativity

waned, with no major works appearing for five years. He worked intermittently and unenthusiastically on a Scribe libretto for the Opéra, *La nonne sanglante*, which was never completed. On the other hand his literary activity was extending beyond the regular demands of his newspaper criticisms to a comprehensive study of orchestration, which began to appear in 1841 in the *Revue et gazette musicale* and which was published as the *Grand traité d'instrumentation et d'orchestration modernes* in 1843.

For the next 20 years much of Berlioz's time was spent on peregrinations of Germany, Austria, Russia, England and elsewhere. Curiosity about advanced music was more evident in such places than in Paris, and the administrative and financial problems of promoting concerts were fewer. The more he travelled the more bitter he became about conditions at home; yet though he contemplated settling abroad – in Dresden, for instance, and in London – he always went back to Paris.

His first concert abroad was on 26 September 1842 in Brussels. His two concerts there were, in Berlioz's words, 'merely an experiment', but sufficiently successful to justify the more ambitious tour that followed shortly afterwards. He was abroad from December 1842 to the end of May 1843, and his tour took in visits to Brussels, Frankfurt, Stuttgart, Hechingen, Mannheim, Weimar, Leipzig, Dresden, back to Leipzig, Brunswick, Hamburg, Berlin, Hanover and Darmstadt. The tour was vividly recounted in open letters to his friends published initially in the *Journal des débats*, then collected in the *Voyage musical en Allemagne et en Italie* in 1844 and finally forming part of his *Mémoires*. He met new friends, including Schumann, revived his

110

acquaintance with old ones – Mendelssohn and Wagner, for example – and made a study of orchestral playing in the different cities he visited. Generally his reception was wholeheartedly warm, a foretaste of many enthusiastic welcomes he was to receive in Germany. To his reputation as a new and original voice as composer was added that of being a leading modern conductor, even though he conducted few works by other composers on his first tour. His return prompted the following reflections:

Paris is where music one moment lies moribund and the next moment seethes with life; where it is sublime and second-rate, lordly and cringing, beggar and king; where it is at once glorified and despised, worshipped and insulted. In Paris music too often speaks to morons, barbarians and the deaf. You see it walking freely and without restraint, or barely able to move for the clammy fetters with which Routine shackles its powerful limbs. In Paris music is a god – so long as only the skinniest sacrifices are required to feed its altars.

Berlioz was accompanied on the German tour by a singer of mixed French and Spanish birth, Marie Recio, who sang in most of his concerts. For her he orchestrated the song *Absence*, from *Les nuits d'été*, which she sang in Dresden for the first time. His feelings for her had none of the passionate élan he had felt towards Harriet Smithson, indeed he tried to escape her pursuit on a number of occasions. On this relationship his letters and writings are more or less silent, yet prosaic or not, it was to last 20 years, until her death. In Paris, Berlioz was now confronted with supporting two households and with the even more distressing spectacle of Harriet Smithson's acute decline. Yet for his son, Louis, Berlioz felt an affection that was to grow stronger until it became the very focus of his emotional life.

The two years that elapsed before undertaking

another concert tour were unremarkable, especially since he was now 40 and *nel mezzo del cammin* of his life. They were not unproductive, but were devoted far more to journalism and publication of his music and of his two first literary works than to composition. The *Mémoires* dwell on his endless obligations as *feuilletoniste* and on his concerts, the largest of which was on 1 August 1844 as part of the Grand Festival de l'Industrie, with over 1000 performers. Four concerts early in 1845 formed a festival promoted by the Théâtre Franconi and were also given with large orchestra and chorus. From this period originates Berlioz's unfortunate reputation as a noisy composer, and the cartoonists were not slow to exploit the image. The finest composition of this period is the *Corsaire* overture, sketched in Nice immediately after the exertions of the Grand Festival de l'Industrie in 1844 and given at first with the title *La tour de Nice*. The broad and majestic *Hymne à la France* also dates from this year. Earlier he had arranged parts of his opera *Benvenuto Cellini* into a brilliant overture, *Le carnaval romain*, played for the first time on 3 February 1844, and an arrangement of Léopold de Meyer's *Marche marocaine* had a notable success a year later. He saw both the *Symphonie fantastique* and the *Symphonie funèbre et triomphale* through the press at this time.

In 1845 began a more intensive and varied succession of concert tours. The first was to Marseilles and Lyons, followed by a visit to Bonn for the Beethoven festival organized by Liszt and attended by leading musicians from all over Europe and a number of crowned heads. There followed a lengthy tour of Austria, Bohemia and

Hungary that brought his name and music even more decisively into the forefront of European attention. Once again he recounted the details of his travels in the *Journal des débats* two years later and subsequently in his *Mémoires*. His itinerary was as follows: by carriage to Linz and thence by steamer to Vienna, where he stayed over two months and gave five concerts. He added two new songs to his repertory, *Zaïde* for soprano and *Le chasseur danois* for bass, and his concerts, which included at least parts of all his major works to date, were a 'grandissime succès'. One concert was devoted to a complete performance of *Roméo et Juliette*, and he had no need to exaggerate his reports of applause and enthusiasm; it was a reception entirely different from anything he had ever experienced in Paris. He then gave three concerts in Prague in as many weeks and then another back in Vienna. In February 1846 he gave three concerts in Pest, including a new arrangement of the Rákóczy March, rapturously received by an audience conscious of its national aspirations. He gave a concert in Breslau, then three more in Prague, where he found the musicians 'generally speaking the finest in Europe' and where he enjoyed success and admiration greater even than in Vienna. On his way back to Paris he gave one concert in Brunswick, on 24 April 1846.

Not only had Berlioz won unprecedented laurels and acclaim on this tour: he had also composed the bulk of a large new work, *La damnation de Faust*. For some years his mind had been turning back to Goethe's *Faust* and the settings he had rejected in 1829. A librettist, Almire Gandonnière, supplied some material before his departure from Paris, and Berlioz wrote the rest himself: henceforth he would write all his own major texts. *La damnation de Faust* was put together in the various cities he stayed in, including Passau, Vienna, Pest, Breslau and Prague. It was completed and orchestrated on his return, although composition was briefly interrupted by the commission of *Le chant des chemins de fer* for the opening of the Chemin de Fer du Nord at Lille on 14 June 1846, an occasion wittily recounted in *Les grotesques de la musique*.

The first performance of *La damnation de Faust*, given on 6 December 1846 at the Opéra-Comique, was a serious reverse, both artistically and financially.

Faust was given twice before a half-empty house. The fashionable Paris audience, the audience which goes to concerts and is supposed to take an interest in music, stayed comfortably at home, as little concerned with my new work as if I had been the obscurest Conservatoire student. . . . Nothing in my career as an artist wounded me more deeply than this unexpected indifference.

Signs of growing philistinism in Paris had been in evidence for some years of course, but the irony was all the sharper in contrast to the warmth and understanding shown to him abroad. Berlioz had no choice but to continue to till foreign soil and to extend the chronicle of his wanderings to new lands. Two principal nations offered hope, Russia and England, and it was to Russia that he went first, within two months of the *Faust* fiasco. Altogether he gave five concerts in St Petersburg and one in Moscow, the former including two complete performances of *Roméo et Juliette*. He now had *La damnation de Faust* to enrich his repertory, and the first two parts were heard three times in Russia. On his way home he gave a complete performance of *Faust* in Berlin

GRANDE SALLE DU GARDE-MEUBLE DE LA COURONNE,
Rue Bergère, n° 2.

Dimanche 24 NOVEMBRE 1839, *à 2 heures précises*,

GRAND CONCERT,

VOCAL ET INSTRUMENTAL,

DONNÉ PAR M.

H. BERLIOZ,

on y entendra, pour la 1re fois,

ROMÉO ET JULIETTE,

SYMPHONIE DRAMATIQUE,

Avec Chœurs, Solos de Chant et Prologue en Récitatif harmonique, composée d'après la Tragédie de *Shakspeare*, par M. H. BERLIOZ. Les paroles sont de M. ÉMILE DESCHAMPS.

PROGRAMME DE LA SYMPHONIE.

N. 1. Introduction instrumentale : { Combats, tumulte. Intervention du Prince. }
1er PROLOGUE (Petit-Chœur.)
Air de Contralto.
Suite du Prologue.
Scherzino vocal pour tenor solo, avec chœur.
Fin du Prologue.

N. 2. Roméo seul. — Bruit lointain de bal et de concert. Grande fête chez Capulet.
Andante et Allegro (orchestre seul).

N. 3. Le jardin de Capulet silencieux et désert.
Les jeunes Capulets, sortant de la fête, passent en chantant des réminiscences de la musique du bal (chœur et orchestre).
Juliette sur le balcon et Roméo dans l'ombre. Adagio (orchestre seul).

N. 4. La reine Mab, ou la fée des Songes.
Scherzo (orchestre seul).

N. 5. 2me PROLOGUE (petit chœur).
Convoi funèbre de Juliette (chœur et orchestre.)
Marche fuguée, alternativement instrumentale et vocale.

N. 6. Roméo au tombeau des Capulets.
Réveil de Juliette (orchestre seul).

N. 7. FINAL chanté par toutes les voix des deux grands chœurs et du petit chœur, et le Père Laurence.
Double chœur des Montagus et des Capulets.
Récitatif, récit mesuré et air du Père Laurence.
Rixe des Capulets et des Montagus dans le cimetière; double chœur.
Invocation du Père Laurence.
Serment de réconciliation; triple chœur.

Contralto solo du Prologue	Mme WIDEMAN.	
Tenor solo du Prologue	M. A. DUPONT.	
Le Père Laurence	M. ALIZARD.	101 VOIX.
Le chœur du Prologue	12 VOIX	
Le chœur des Capulets	42 VOIX	
Le chœur des Montagus	44 VOIX.	
Orchestre	100 INSTRUMENTS.	

L'exécution sera dirigée par M. H. BERLIOZ.
Maître de chant : Mr DIETSCH.

Dimanche 1er Décembre 2me Concert (Roméo et Juliette).

PRIX DES PLACES : 1res Loges, 10 f.; Stalles de Balcon, 10 f.; Secondes Loges, 6 f.; Stalles d'Orchestre, 6 f.; Loges du Rez-de-Chaussée, 6 f.; Parterre 3 f.; Amphithéâtre, 2 f.

On trouve des Billets chez M. RÉTY, au Conservatoire; et chez M. SCHLESINGER, rue Richelieu, 97

Imprimerie de VINCHON, rue J.-J. Rousseau, 8.

---

11. Programme of the first performance (24 November 1839) of Berlioz's 'Roméo et Juliette'

115

at the invitation of the King of Prussia. Once again he was able to report, on his return:

great success, great profit, great performances, etc. etc. . . . France is becoming more and more philistine towards music, and the more I see of foreign lands the less I love my own. Art, in France, is dead; so I must go where it is still to be found. In England apparently there has been a real revolution in the musical consciousness of the nation in the last ten years. We shall see.

So he left Paris once again, reaching London in early November 1847. He had been engaged by Louis Jullien as conductor of the opening season at Drury Lane Theatre, and the works in his charge were Donizetti's *Lucia di Lammermoor* and *Linda di Chamonix*, Balfe's *The Maid of Honour* and Mozart's *Le nozze di Figaro*. The season opened in December, yet within a month Berlioz was sensing alarm at Jullien's approaching bankruptcy. Jullien had all the bravado and showmanship of the charlatan, and though the opera season ran its full two months Berlioz was never paid. He pinned his hopes, instead, on a concert of his own music, which won many admirers. At the same juncture revolution broke out in Paris, and Berlioz was perhaps thankful to be away from the barricades. He began to piece together his *Mémoires* and added a preface that despairs of artistic life in France. His one salaried post, as librarian of the Conservatoire, was threatened, and many of his friends were fleeing the Continent to settle in England, chief of them Charles Hallé. Despite Jullien's failure Berlioz found the English friendly and hospitable and their appetite for music encouraging. A second major concert on 29 June 1848 in Hanover Square Rooms established his reputation, especially in the eyes of the London press, and he contemplated staying if a suitable position were offered to him. Yet he returned to Paris, perhaps because his *feuilletons* offered him his sole regular income and because it was after all, as he himself ironically noted, his home.

V 1848–63

From this time forward Berlioz's tours to foreign cities were almost all to places he had visited before; his years of first conquest were over. In the space of six years his European fame had flowered and he had, too, published most of his major works (*Benvenuto Cellini* and *Faust* were exceptions), making possible the further dissemination of his music. Success abroad went a long way to compensating for failure at home, and he continued to make regular visits to England and Germany for 15 years. The new regime in France made the Romantic heyday seem even more remote, and soon Second Empire tastes were to infiltrate all walks of life. But Berlioz achieved a new lofty detachment based on his powerfully ironic sense of humour and on his deep-rooted faith in classical ideals. One may detect a new repose in his music after 1850, linking him with his adored Gluck and isolating him both from Parisian taste and from the new schools of Liszt and Wagner.

It is not necessary to chronicle every foreign tour of this period. The majority gave him deep satisfaction and showed a genuine understanding in his audiences. The most notable events were in Weimar where Liszt's position at the Duke of Saxe-Weimar's court allowed him to further his admiration of Berlioz. In March 1852 Liszt revived *Benvenuto Cellini* for which Berlioz, with Liszt's aid, devised a new version, partly out of despair at its unhappy fate in 1838 and partly to meet the demands of

German taste. Its success in Weimar and subsequently in other German cities was lasting. In November 1852 Liszt gave a Berlioz Week, with *Benvenuto Cellini* and *Roméo et Juliette* and two parts of *La damnation de Faust*, which was later dedicated to Liszt when published in 1854. In reply Liszt dedicated his own *Faust Symphony* to Berlioz. Further visits by Berlioz to Weimar in 1855 and 1856 were the occasion of discussions in which the Princess Sayn-Wittgenstein, Liszt's mistress, urged Berlioz to pursue his dream of a large epic opera based on the *Aeneid*; it came to fruition in *Les troyens* in 1858.

But in London *Benvenuto Cellini* fared badly when it was performed there, at Covent Garden in June 1853, and was rapidly withdrawn, as it had been in Paris in 1838. This was a single blot on Berlioz's otherwise happy reception in England in all his five visits. His stay in 1851, as a member of the international jury to examine musical instruments at the Great Exhibition, produced some remarkable impressions in his reports to the Paris press, above all the experience of hearing 6500 children intoning *All people that on earth do dwell* during the Charity Children's annual service in St Paul's. Six concerts in Exeter Hall in 1852, in which his own music had relatively little prominence, were 'an altogether extraordinary success exceeding anything I had had in Russia and Germany'. Two performances of Beethoven's Choral Symphony, in particular, set the seal on his celebrity as a conductor, and contributed to an invitation from the New Philharmonic Society to conduct their 1855 season. Since Wagner was then conducting the old Philharmonic Society it provided an occasion for the two men to meet and to exchange sympathy and encouragement in fuller measure than at any other time in their careers. Subsequently their radically divergent conceptions of music were to bring an estrangement between them. Other foreign visits that Berlioz recalled with satisfaction were to Hanover, Brunswick and Dresden in 1854, to Brussels in 1855, and his regular engagements for the summer season at Baden-Baden. He first conducted there in 1853 and was engaged every year from 1856 to 1863. Bénazet, manager of the casino, 'let me have everything I could possibly want for the performance of my works. His munificence in this respect has far surpassed anything ever done for me even by those European sovereigns whom I have most reason to be grateful to'. It was Bénazet who commissioned for the Baden-Baden theatre Berlioz's last work, *Béatrice et Bénédict*, first performed in 1862.

At home in Paris Berlioz made another determined attempt to win an audience for his music by the formation of a Société Philharmonique, in clear rivalry to the Société des Concerts du Conservatoire. This new body gave its first concert on 19 February 1850 with Berlioz as conductor. Despite initial success the society was troubled by internal dissent and by an early shortage of funds, and lasted only until May 1851. But in that period Berlioz had conducted a wide range of music and had introduced some of his own works, notably *L'adieu des bergers*, later to be the central part of *L'enfance du Christ*. At its first performance Berlioz attributed it to an imaginary 17th-century composer Pierre Ducré, allowing him to delight in the delusion of his audience. The Société Philharmonique also gave his Requiem in the church of St Eustache. The complete *L'enfance du Christ* was first heard in Paris on 10 December 1854 having grown from *L'adieu des bergers* and *La fuite en Egypte*. Many critics observed a more restrained style in the work, but Berlioz insisted that on the contrary only his subject matter had changed and that his primary stylistic aim, accuracy of expressive content, was still unchanged.

Berlioz's monumental manner was represented by the *Te Deum*, composed in 1849, although the conception probably goes back three or four years earlier. He found no opportunity to perform this work until April 1855, when it was included in the large-scale events promoted in connection with the Exposition Universelle. By that time he had added to its two choruses a part for large children's choir, inspired by his experience in St Paul's Cathedral in 1851. In November 1855 Berlioz conducted three big concerts in the Palais de l'Industrie for the closing events of the Exposition. Throughout the 1840s and most of the 1850s the Société des Concerts, Paris's longest-established and most regular concert-giving body, continued to ignore Berlioz's music completely.

His father died in 1848. Berlioz had felt deeply attached to him, the more since the strain in their relationship during Berlioz's first days in Paris had passed, and he felt the loss keenly. He remained close to both his surviving sisters, Nanci and Adèle, who died in 1850 and 1860 respectively – and their families. He inherited a modest income from his father's estate, which relieved some of his financial burdens. Harriet Smithson died in 1854 after four years of severe paralysis. Berlioz wrote movingly of her and of the failure of their happiness in the *Mémoires*; he never forgot the impression she first made on him or the style of dramatic interpretation that coloured his own conception of Shakespeare.

120

He married Marie Recio seven months later, a natural step after their 12-year association, and though she had not sung in public for some years he still had to suffer the damage done by her spiteful attitude to other musicians, Wagner especially. With her came her Spanish mother who outlived them both and cared generously for Berlioz in his last years. His son Louis, now in the French navy, caused Berlioz many an anxiety after a difficult adolescence, but gradually there developed a strong bond between them. In Louis' words: 'The thread of my life is but the extensions of my father's. When it is cut, both lives will end'. Louis saw action in the Crimean War and in the Baltic. In 1867, when captain of a merchant ship on the French expedition to Mexico, he died of yellow fever in Havana, one of the severest blows Berlioz ever had to suffer and a direct contribution to his own final decline. In Louis' love of travel and the sea Berlioz saw a reflection of his own lifelong, idealized passion for distant lands, inextricably interwoven with his dream of a land where art and music enjoyed unfettered cultivation, where the frustrations and miseries of Paris were not to be found. In 1862, in response, Louis came to love and admire his father's music.

Berlioz's compositions in the 1840s were haphazard in origin and frequency, partly because of his diversion of energy to travel, conducting, proof correction and journalism. In the following decade these diversions were no less pressing but he now found the mental and spiritual calm to produce a series of masterpieces that shine nobly through the day-to-day battles he was obliged to fight. After the *Te Deum* of 1849, his main productions were *L'enfance du Christ*, composed to his own text mostly in 1854. Another work of 1854 is the occasional and unimportant cantata *L'impériale*, for the Exposition Universelle. Early in 1856 he orchestrated most of *Les nuits d'été* (*Absence* had been orchestrated in 1843) for publication in Winterthur, though he never heard more than *Absence* and *Le spectre de la rose* in orchestral form. At that point (April 1856) he yielded to his desire to compose a vast epic opera based on the second and fourth books of Virgil's *Aeneid*. The idea had been in his mind for five years or so, and had doubtless haunted him since childhood when he wept at his father's readings of Virgil. His love of Virgil had stayed with him even through the blinding discoveries of Shakespeare and Goethe and came back to him with irresistible force in his maturity. His muse was now in full flight, for long disillusionment with the world seems to have fanned the creative flame even though he knew what difficulties he would face if it were ever written. By abandoning most of his concert tours and much of his journalism he did in fact complete *Les troyens*, words and music, in less than two years, with small additions and revisions to be made at intervals over the next five. It is a five-act grand opera in the French classical tradition, on the same approximate scale as Meyerbeer's operas and many others that enjoyed regular performance in Paris. Yet Berlioz's chances of securing a production in which his work would receive attention at all close to its merits were negligible from the first – a fact he was fully aware of.

The following five years were devoted to a series of frustrating attempts to see *Les troyens* on the stage. Berlioz's enemies in the press were quick to exaggerate its length and its demands, and the failure of *Benvenuto Cellini* was still remembered at the Opéra. He gave numerous readings of the poem to carefully chosen audiences; he vainly sought the patronage of Napoleon III and his ministers. Eventually, in 1860, he accepted an offer to mount it at the Théâtre-Lyrique, an independent theatre run by the enterprising impresario Carvalho, while Wagner's *Tannhäuser* was staged with unprecedented extravagance at the Opéra. Its failure in March 1861 was bitterly ironic for Berlioz, and it created an opportunity for *Les troyens* to be accepted at the Opéra. This agreement fell through early in 1863 so turning Berlioz back to the Théâtre-Lyrique, where, in order to see any production at all, he was forced to divide his opera into two parts, Acts 1 and 2 becoming *La prise de Troie* and Acts 3 to 5 *Les troyens à Carthage*. The second part was first performed on 4 November 1863, with Mme Charton-Demeur as Dido. It was an unequivocal success, warmly admired by the majority of the press and running to 21 performances. Berlioz was proud and touched, but gradually embittered, then enraged, to see cuts made by Carvalho at subsequent performances (the 'Chasse royale et orage', for example, was played on the first night only) and to see the printed vocal scores being mutilated to match the performances 'like the carcass of a calf on a butcher's stall'. Of *La prise de Troie* Berlioz only ever heard one extract sung at Baden-Baden in 1859.

VI 1863–9

After the year 1863 Berlioz discouraged revivals of *Les troyens* and none took place for nearly 30 years. The financial fruits were compensation for his artistic despair, for he was enabled at long last to resign his duties as critic of the *Journal des débats*. He retired from composition and criticism, and allowed his spirit to be overcome by a despair and disillusionment of appalling intensity. He became morbidly conscious of death, especially since the loss of two sisters and two wives, and as more and more of his contemporaries and friends disappeared he haunted the cemeteries. In 1864 he wrote:

I am in my 61st year; past hopes, past illusions, past high thoughts and lofty conceptions. My son is almost always far away from me. I am alone. My contempt for the folly and baseness of mankind, my hatred of its atrocious cruelty, have never been so intense. And I say hourly to Death: 'When you will'. Why does he delay?

And yet he lived another five years, suffering acutely from a form of intestinal neuralgia that had first appeared some ten years before and had reached severe proportions by 1859. Physical pain was never far away in the last 15 years, accentuated by his spiritual isolation. He depended more and more on a diminishing circle of friends for comfort, especially Stephen Heller, the Damckes, the Massarts and Edouard Alexandre. From time to time he would give readings of Shakespeare; but music he usually avoided. He went to few public concerts or operas, making an exception for *Don Giovanni*, for Pasdeloup's concerts where *Les francs-juges* overture and parts of *Les troyens* were played, and for the Opéra's revival of Gluck's *Alceste* in October 1866, which he was asked to supervise. He completed and revised his *Mémoires*. 1200 copies were printed in 1865 and stored in his office in the Conservatoire. A few close friends received copies; the rest were to be published after his death.

The final pages of the *Mémoires* reveal the single ray of light that penetrated an otherwise all-pervading

a Madame de Milde
un admirateur de son talent
un adorateur de sa grâce et de sa beauté

Hector Berlioz

12. *Hector Berlioz: photograph, c1865*

gloom. In 1864 he felt an overwhelming impulse to revisit the scenes of his childhood, especially Meylan, near Grenoble, where his adored Estelle had lived as a child. He had made an earlier pilgrimage and even written to her in 1848, but this time, having discovered that she was living in Lyons, he wrote again and paid her a visit. She was now a widow of 67, he 60, yet the memory of their childhood encounter was fully alive in his mind. 'My soul leapt out towards its idol the moment I saw her, as if she had still been in the splendour of her beauty.' Berlioz was enraptured to be in her presence, to kiss her hand, and, next day, to receive even a brief and formal letter from her. He sought permission to write to her, and for the rest of his life he did, nearly every month. He visited her the following three summers in Geneva, where she went to live with her son. She accepted his attentions with calmness and incomprehension turning gradually to understanding and sympathy. The full extent of his dependence on this glimpse of his own childhood cannot be measured: not for the first time he had fallen in love with an idealized vision, reality transfigured by imagination.

Berlioz had not wholly given up conducting his own music abroad. In December 1866 he accepted an invitation to conduct *La damnation de Faust* in Vienna. Hanslick castigated the music but in general its success was immense. Age, illness and his poor knowledge of German now impaired his conducting skill, but he was lionized by Cornelius and Herbeck and fêted as he had been in 1845. The following February he conducted *Harold en Italie* and parts of *Béatrice et Bénédict* in Cologne as the guest of his old friend from 1830, Ferdinand Hiller. The final burst of energy was his acceptance of an invitation to St Petersburg in November 1867, shortly after the death of Louis. Perhaps he thought he would find renewal and escape. Instead the journey and the concerts – six in St Petersburg and two in Moscow – shattered his remaining strength. Not even the instinctively sympathetic response from the emerging school of Russian composers or the overwhelming public applause staved off a sense of impending collapse. He went directly to Nice, scene of happy memories of 1831 and 1844, and Monte Carlo. Twice, walking by the sea, he fell and was picked up dazed and bleeding. He returned to Paris where he had 12 months to live, now little more than a shadow, dragging out what had come to seem a meaningless existence. He died on 8 March 1869, having been cared for by his mother-in-law and visited by his remaining friends, the Damckes, Saint-Saëns and Reyer. He was buried in the Cimetière Montmartre on 11 March 1869.

CHAPTER TWO

Character and personality

Berlioz was widely misunderstood in his lifetime – and has been since – despite the clarity of his ideas and the abundance of his writings. Few composers have explained at such length or with such cogency the nature of their inspiration, its sources, its aims or its meaning. In Berlioz certain qualities stand out, and chief among them must be cited his consistency and his sincerity. There was no dividing line between his life and his music; the same principles governed both and each was a reflection of the other. Few composers have woven their own personality so tightly into their music, so that all his works reflect something in himself expressed through poetry, literature, religion or drama. Expression is the key. Music was not for him an autonomous art obeying internal rules and exploiting internal relations. It was an integral part of emotional and spiritual life, reflecting the teeming motion of the mind, the explosive diversity of life. Just as Shakespeare had laid bare every facet of human nature in poetry and drama, so Berlioz aspired to chart his own experience in music. Of course, not all his music is autobiographical, but he remained steadfast to an ideal of truthfulness of artistic expression from the first note he wrote to the last.

Sincerity as burning as this meant an unwillingness to compromise. He was not a diplomat, and he often failed to win influential friends by being outspoken at unguarded moments. Cherubini and Fétis, for example, might have been won over to his cause had he been less severe on their work to their faces. Many acquaintances found him embarrassingly forthright in his views. 'Few were at ease in his company', wrote Legouvé; and his sister Nanci said as early as 1824 that with him everything was open and spontaneous and that he never made the slightest effort to conceal the vagaries of his mood. He was subject to violent emotional change, from enthusiasm to misery, and these are reflected in his music. The *Symphonie fantastique* specifically depicts the *vague des passions* of the young artist. Enthusiasm was to be seen in his adoration of Harriet Smithson, Shakespeare, Gluck, Goethe and Virgil; misery in his descriptions of the 'spleen' and the 'mal d'isolement' that afflicted him throughout his life, increasingly so as his isolation became more real and more intense. His dislikes, for example of inexpressive music, ornamented singing and of commercially minded theatre managers, were as intense as his enthusiasms and as consistently articulated. Sometimes he had to conceal his feelings, when writing public notices of works by respected contemporaries, a cause of bitterness about the critic's métier that gradually intensified.

Another cause was the failure of the world to live up to his ideals, and not just its failure, its clear determination to dissociate itself from them. He was a passionate idealist, whose conception of what might be achieved in music drew him on even when external discouragement was most intense: in the treatise on orchestration he described an ideal orchestra; in *Les soirées de l'orchestre* he described an ideal city, Euphonia, where everything is arranged to the service of art and where commerce has no place. He expected his audience to have an imagination as vivid as his own, he made no rigid frontier between the kingdoms of imagination and reality. His lofty conception of the role of art and music presupposed an essentially aristocratic view that music was not for the many; it was a highly sophisticated form of expression (sometimes disarmingly simple in its outward form) that required the highest degree of imagination and intellect for its proper appreciation. He spoke as an artist to men of kindred capacity. The morons, fools, parasites and tune-mongers who made fortunes out of music he bitterly despised.

One quality saved him from morbid self-pity and from the tediousness of rapture: he had the sharpest sense of the ridiculous in human behaviour. His writings win us by their humour as much as by their style and their ideas. His conversation was laced with puns. His humour was largely based on the ironic, on the startling contrast between what is and what might be, but also on the foibles of singers, pianists and audiences. Contrast and diversity were to be cultivated as well as to be laughed at, so that the juxtaposition of passion and mockery in his writing is equally as characteristic as the simultaneous combination of opposites to be found so often in his music.

His intellect embraced the broad movement of ideas that Romanticism swept into currency and can be seen as a powerful expression of them. Yet some tastes passed him by, notably the nostalgia for things medieval so widespread in his time. He had little taste for painting and had, as he confessed, 'little feeling for conventional beauty'. He knew classical Latin and French literature well and was given to quoting it at all times, especially to adapting quotations slightly to suit his purpose. His favourite authors remained with him through life and provided imaginative worlds for his fantasy. Travel books likewise absorbed him.

Berlioz was no philosopher, since life, for all his idealism, was a practical matter whose problems had to be confronted by action not theories. The business of composing and performing music, educating listeners and guiding taste was a daily obligation from which he drew such evidence as he needed to reinforce his views on the proper place of music in culture. His historical sense did not extend to an appreciation of music much earlier than that of Gluck, and most of Bach and Handel left him cold.

Although he falls clearly into the French tradition, from Rameau and Gluck through Gossec, Méhul and Le Sueur, he had no nationalist preconceptions whatever. His distaste for Italian music stemmed purely from its composers' higher regard for melody and vocalization than for expression. Because Germany, England, Bohemia and Russia applauded his music he was prepared to regard these as musical nations without going so far as to presume that their compositions were inherently superior to those of any other country. He was as happy to work with foreign musicians as with French, though he never fully mastered English and spoke little German. Many of the friends he valued most highly were foreigners, such as Hiller, Heller, Ernst, Davison, Damcke, Hallé and of course Liszt.

Slowly the youthful idealist changed into the aloof, dignified but weary figure of later years. His working life was characterized by tireless energy and the capacity to turn emotional stress into creative form. He

131

despised his enemies and this was all part of the intensity of his emotional being, with nerves that responded more sharply than those of his fellow men. His life was a continuous search for an unattainable tranquillity, not the tranquillity of idleness or repose but the peace of mind that would allow him to work rather than labour, write music rather than prose, and grapple with his leaping imagination rather than with the petty squabbles of everyday life.

CHAPTER THREE

Works

A fundamental factor in the understanding of Berlioz's music is to recognize that for him there existed rigid categories of neither form nor medium. The genres of opera, cantata, song and symphony all merge imperceptibly one into another and overlap constantly. The important criterion is the matching of means to expressive ends. Heterogeneous elements are to be found in all his large-scale compositions and some, like *Lélio* or the *Huit scènes de Faust*, have unity of subject matter and artistic purpose, not unity of musical means. He did not refuse to adopt conventional means in order to be iconoclastic; he felt impelled to give every idea its proper musical setting according to its literary, pictorial or suggestive content, and this led him to construct new forms and to throw musical genres into new relationships.

Apart from two youthful quintets and a sextet, all now lost, Berlioz composed no chamber music. He wrote nothing for the most widely cultivated instrument of his time, the solo piano. He was not a pianist and his only keyboard music is a group of three short harmonium pieces commissioned in 1844. His chosen medium was the orchestra, in his time expanding with new speed and momentum. The best of his songs, though all composed with piano accompaniment, were eventually orchestrated. He was not, like Chopin or

Schumann, a miniaturist by habit, yet the smallest of his works are little more than albumleaves, while the largest are conceived on a huge scale.

I Operas
Berlioz completed five operas, all of which differ in style and dramatic stance. He contemplated or sketched many more and had at least one operatic project in mind at nearly all points of his working life. Opera was the medium of the predecessors he admired most – Gluck, Spontini, Méhul, Le Sueur, Weber – and was the most assiduously cultivated form of music in Paris during his first years there. Success in opera was also, in Berlioz's time, the principal yardstick by which a composer was measured and the surest way to financial reward. Dramatic expression is the very pulse of his music, so that operatic elements are to be observed in many of his non-operatic works, especially the symphonies and larger choral works.

Estelle et Némorin, composed in 1823, has not survived. There followed *Les francs-juges*, composed to a libretto by Humbert Ferrand in 1826. The secret tribunals of the Black Forest in the later Middle Ages provided a background for a sombre story of heroism and virtue in the face of oppression and tyranny. Its colour came from Méhul and more especially Weber, whose *Freischütz* had been heard in Paris two years before. Six complete numbers and the overture survive, of which the latter is a bold and imaginative piece of orchestral writing, especially since Beethoven was still unknown to him. The first version, probably consisting of 14 numbers, was superseded in 1829 by a longer version, which probably included the *Marche des gardes*, later to become the 'Marche au supplice' in the *Symphonie fantastique*. But though Berlioz was soon pillaging the score, he made a further attempt to recast it in 1833 into a one-act intermezzo *Le cri de guerre de Brisgau*. His failure with this reflects the enormous strides his music was taking during these years and his desire to tackle new material. Consider his remark in 1828 that he had two operas in hand for the Opéra-Comique, a third for the Opéra and a fourth planned on the English play *Virginius*, by J. S. Knowles. Projects on *Robin Hood* and *Atala* perhaps never even reached libretto stage. In 1833 he considered, *Much Ado About Nothing* for the Opéra-Comique, the theatre for which *Benvenuto Cellini* was destined when first drafted.

This opera, when finally presented in 1838, was utterly different from *Les francs-juges* in pace, colour, subject matter and dramaturgy, and was very different too from any other opera to be seen in Paris in that decade, comic or serious. It combined elements of both, perhaps accidentally, because the work, originally intended for the Opéra-Comique, had been upgraded for the Opéra. Yet the mixture of genres was utterly characteristic of Berlioz, with the tone veering from knock-about comedy to serious reflection on the duties and priorities of the artist. One reason for its poor reception was the dazzling brilliance of the music, its orchestral virtuosity and rhythmic élan, shifting in metre and colour with kaleidoscopic suddenness. Neither players, singers nor audience could grasp an opera so teeming with life when the more stolid manner of Meyerbeer was fashionable. The libretto has weaknesses, especially in the character of the Pope (changed to Cardinal on the order of the censor), and is diffuse in the last act; it was

13. Autograph MS (with the composer's revisions) of part of the 'Royal Hunt and Storm' from Berlioz's 'Les Troyens', composed 1856–8

these problems that Berlioz attempted to solve in 1852 when he recast the work for Weimar. But at the same time other dramatic inconsistencies were created and some of the vitality of the 1838 version was sacrificed. In his own words, *Benvenuto Cellini* 'contains a variety of ideas, an energy and exuberance and a brilliance of colour such as I may perhaps never find again'.

Berlioz made some sketches on Ballanche's *Erigone*, an 'intermède antique' in the period 1838–40; he also considered a collaboration with Frédéric Soulié, and finally began work on *La nonne sanglante* by Scribe, a concession to the Opéra's established tastes. The libretto, based on Lewis's *The Monk*, recalls the sombre tones of *Les francs-juges* and the music survives, likewise, only in fragmentary form, for Berlioz abandoned composition and negotiation with the Opéra in 1847. What music survives is undistinguished, hampered by Scribe's lumbering metres, though sometimes prophetic of the restrained accents of *Les troyens*.

By 1850 Berlioz seems for the first time to have abandoned thoughts of opera. Nonetheless, involuntarily, the dream of a grand opera on Virgil's *Aeneid* began then to impose itself. He resisted it, but in 1856, urged by the Princess Sayn-Wittgenstein, decided to yield. He wrote his own libretto, building Acts 1 and 2 around the tragedy of Cassandra in Troy and Acts 3, 4 and 5 around the tragedy of Dido in Carthage, linked in the character of Aeneas and the fateful destiny of the Trojan people. It is a truly epic opera, grand in conception and execution, with equal claim to be Berlioz's masterpiece as to be one of the towering achievements of 19th-century music. In it all aspects of Berlioz's art converge: the monumental and the intimate, the symphonic and the operatic, the decorative and the solemn. Its great scenes include the enormous finale of Act 1 where Cassandra's wails of doom contrast starkly with the Trojans' fatal faith in the Wooden Horse; the Royal Hunt and Storm, where the coming together of Dido and Aeneas is enacted in an elaborate and symbolic mime; the sublime sequence of quintet, septet and duet in the garden scene that follows; the final departure of Aeneas and Dido's immolation, and much else. The opera belongs to a long tradition that embraces Rameau, Gluck, Spontini and Meyerbeer, and was anything but revolutionary. Yet the classical poise and sense of tragedy is imbued with a warmth of feeling and passion that only a Romantic composer could attain, 'Virgil Shakespeareanized', he called it.

His last opera was an *opéra comique*, composed almost as relaxation after the travail of *Les troyens*. *Béatrice et Bénédict* was begun in 1860 and first performed in 1862. Berlioz made his own libretto from Shakespeare's *Much Ado About Nothing*, adapting much of the original text for his dialogue. 'It is a caprice written with a point of a needle', in his own words, a fair description of the light textures and disarming immediacy of the work. There are moments of sterner feeling, as in Beatrice's air in Act 2, ensembles of almost Mozartian grace, and a heaven-sent tranquillity in the Duo–Nocturne that concludes the first act. As an interpolation on Shakespeare, Berlioz invented a comic character Somarone, in which the archetypal pedantic Kapellmeister is gently satirized.

II Symphonies
Berlioz's discovery of Beethoven led him to compose symphonies, yet his treatment of music as an expressive

and dramatic art made them into something other than the pure instrumental music that many Germans saw in Beethoven. They stretch the meaning of the word to new limits. The first, the *Symphonie fantastique* of 1830, is a five-movement symphony with a slow introduction to the first-movement Allegro, a waltz, a slow movement, a march and a finale, the whole unified by a theme that recurs, transformed, in each movement. But it is, more importantly, an 'Episode in an Artist's Life', set out in detail in the programme, and the recurrent theme is an *idée fixe* representing the artist's obsession with the woman he adores. There is no mistaking the artist or the woman as Berlioz and Harriet Smithson, and the programme spells out his dreams and fantasies in dramatic form. The slow introduction, for example, portrays the 'flux of passion, the unaccountable joys and sorrows he experienced before he saw his beloved'; the Allegro describes 'the volcanic love that his beloved suddenly inspired in him'. The last two movements represent an opium dream in which he dreams he has murdered his beloved and is led to execution, and in the finale he finds himself at a macabre and turbulent witches' sabbath. Later (in 1855) the programme was altered to interpret the whole drama as a dream, not just the end. Berlioz devoted much time and attention to the programme, revised it frequently and generally issued it as a pamphlet when the symphony was performed. Its vivid action is matched by music of unprecedented boldness and originality. The orchestration adopts many practices previously associated with opera, such as the use of harps, bells and english horn. Berlioz used the E♭ clarinet for the shrieking presentation of the beloved's image in the finale, and brought together combinations and distributions (for example the multi-divisi strings) of extreme boldness; four timpani are used simultaneously to represent distant thunder, and the brass is given a distinctive new role. The novelty and defiant youthfulness of the score have never faded and the musical and thematic invention is inextricably linked with Berlioz's conception of a new world of colour and dramatic content. At one stroke the symphony as a form became a fully-fledged medium of explicit drama.

Harold en Italie, the symphony that followed in 1834, has a prominent concerto element, with a solo viola impersonating the character of Harold, a responsive and passionate observer of scenes of Italian life. The drama is more episodic and less cogent than in the *Symphonie fantastique* and the *idée fixe* that here represents Harold recurs unchanged in each movement. There is a direct link with Beethoven in the last movement ('Orgie de brigands'), which is introduced by brief reminiscences of the first three movements. There are picturesque echoes of Italy in the 'Marche des pèlerins', with its tolling bells and chanting pilgrims, and the serenade of the Abruzzi mountaineer. The symphonic idea is retained with limited acceptance of the principles of sonata form in the first movement and by the balance of the four movements. The music is also enriched by an obsession with rhythmic vitality and rhythmic experiment, looking forward to the impulsive vivacity of *Benvenuto Cellini*.

Berlioz's third symphony was *Romeo et Juliette* (1839), sub-titled 'symphonie dramatique'. It moves well away from the purely symphonic realm towards that of opera. Yet Berlioz was specifically not writing an opera, and he kept the idea of symphonic construction

140

closely in mind. He was able, consequently, to express the main portions of the drama in instrumental music, while setting the more expository and narrative sections for voices. The three principal instrumental sections – 'Fête chez Capulet', 'Scène d'amour' and 'La reine Mab' – can be seen as first movement, slow movement and scherzo, with elaborate vocal introduction and finale. The introduction sets the scene with warring Montagues and Capulets, and outlines the coming drama in choral recitative, with foretastes of later movements and solo sections for tenor and contralto. The text is by Emile Deschamps, based on Garrick's version of Shakespeare (which is what Berlioz saw at the Odéon in 1827). The instrumental sections intensify the drama, since instruments have a more powerful capacity for deep expression than voices, as Berlioz explained in his preface. The finale is a complex sequence of movements, scarcely symphonic in the traditional sense, but drawing the listener out from the inner drama to the world of action and resolution. Juliet's funeral procession, the scene where Romeo comes to the vault, the death of the lovers, Friar Laurence's explanation and the reconcilia-tion of the two families are enacted in music mostly of operatic cast, especially the final Oath, which can match anything in *Guillaume Tell* or *Les Huguenots* for gran-deur, and was later echoed in *Tannhäuser*.

The *Grande symphonie funèbre et triomphale* (1840) is an occasional piece for a solemn public ceremony, scored for large military band, and was probably put together from earlier drafts. This is definitely so for the second movement 'Oraison funèbre', reworked as a trom-bone solo from a scene in *Les francs-juges*. The first movement, 'Marche funèbre', is one of Berlioz's most

141

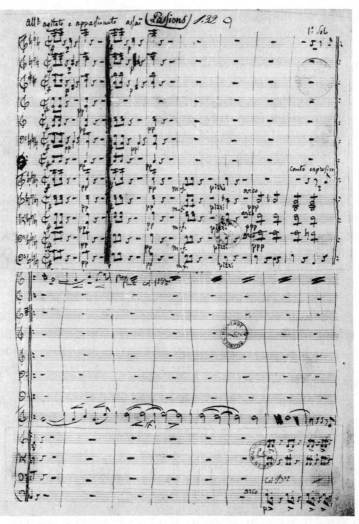

14. *Autograph MS of the passage containing the beginning of the idée fixe from Berlioz's 'Symphonie fantastique', composed 1830*

powerful movements, immense in span and dynamic contrast, with an overwhelming sense of melancholy projected on to a public, even popular level. The finale, 'Apothéose', is a triumphal march. Of particular interest is the fact that the three movements are in different keys, F minor, G major and B♭ respectively. In 1842 Berlioz added parts for optional string orchestra and later also for chorus, with a patriotic text by Antony Deschamps.

In his *Mémoires* Berlioz spoke of a symphony he dreamed of writing in the 1850s, though because of difficulties of time, expense and performance he decided not to commit it to paper, a tragic loss perhaps, but an indication too that he had still not lost sight of the symphonic mode. Yet it is easier to see the symphonic impulse expanding vastly into *Roméo et Juliette* and thence further towards operatic expression in *La damnation de Faust* and *L'enfance du Christ*.

III Choral works
Berlioz described *La damnation de Faust* as an 'opéra de concert' at its time of composition, but it was finally issued as a 'légende dramatique'. In 1847, when there was a proposal to turn it into an opera, it became clear that he would have wanted to revise it considerably for the stage. Its effect rests too strongly on the imagination to be directly transferable to the theatre, and the same can be said of *Roméo et Juliette* and *L'enfance du Christ*. Transformations of time and place are sometimes dramatically sequential and sometimes kaleidoscopic, since Berlioz used only those parts of Goethe's *Faust* that met his needs. Taking the rejected *Huit scènes de Faust* of 1828–9 and inserting his rousing arrangement of the Rákóczy March at the end of the

143

first part, he expanded the work into a broad conception of Faust as an aspiring, yearning soul, overwhelmed by the immensity of nature, with a heart sensitive to emotion at many levels, yet ultimately damned by his inner weaknesses, which Mephistopheles both represents and exploits. The nature music is particularly striking, in Faust's welcome of spring at the beginning and the invocation early in the fourth part, where harmony and orchestration display Berlioz's genius for the unexpected within the span of a huge melodic line. The chorus plays a large part, as penitents, carousers, sylphs, soldiers, students and as the occupants of both Heaven and Hell. The finale of the second part, combining the songs of both soldiers and students, is a tour de force; and the Pandaemonium, at the climax of the precipitous 'Course à l'abîme', is an apocalyptic scene worthy of John Martin (a comparison first made of Berlioz's music by Heine) or even Blake.

L'enfance du Christ (1850–54) shows the same mixture of dramatic action and philosophic reflection as *La damnation de Faust*, though Berlioz still refrained from calling it an oratorio. It is constructed in three parts, 'Le songe d'Hérode', 'La fuite en Egypte' and 'L'arrivée à Saïs', the second of which was composed first. Like *La damnation de Faust*, the score contains stage directions to explain (to the imagination) the movement of events. The third part, with the Ishmaelites' welcome of the holy family, is the most immediately theatrical. In the first part Berlioz's concern was for the tormented soul of Herod, disturbed in his dreams yet at the mercy of his soothsayers; then, with a clear change of mood, the listener is taken to Mary and Joseph in Bethlehem and the angels' warning. The second part is seen largely through the eyes of the narrator, with instrumental music in the overture setting the tone and distancing the action. At the end, when the Saviour has found repose, the music draws away from the portrayal of action to a serenely contemplative farewell, 'O mon âme', the nearest Berlioz ever came to a devoutly Christian mode of expression.

Perhaps these dramatic choral works would never have existed if Berlioz had won early success and acceptance in opera. Yet they constitute a heterogeneous genre entirely characteristic of his faith in expressive truth as superior to consistency of method. They left their mark, too, on the dramatic style of *Les troyens* into which his symphonic, choral and dramatic impulses were then all compulsively channelled.

Berlioz was not an orthodox Christian, yet he set sacred texts with a strong personal vision that has deep religious roots. The Requiem (1837) and *Te Deum* (1849) form a pair of monumental works exploiting his sense of numinous space on a grand scale. Space and direction are essential elements in both. In the Requiem the large orchestra is supplemented by eight pairs of timpani and four groups of additional brass placed at the four corners of chorus and orchestra. These large forces are used for the 'Tuba mirum', where Berlioz's vision of the Last Judgment is realized with overwhelming vividness and force, and there is no doubt that the music requires a building (such as the church of Les Invalides, for which it was composed) that can do justice to its sonority. This broad ceremonial style was a legacy from the outdoor music of the French Revolution, when immense forces of wind and percussion were assembled for public occasions; yet Berlioz was careful to contrast the great

145

with the small. The 'Quid sum miser' and the 'Quaerens me' form a strikingly restrained contrast with the outbursts on either side of them. The Offertorium is written in a subdued contrapuntal style, with the chorus intoning two alternating notes over a winding orchestral accompaniment. The Sanctus is a trifle worldly in its sweetness, and the 'Hostias' exploits the extraordinary sonority of high flutes and low trombones in combination. The Requiem is expressive without being theatrical, solemn without being sanctimonious. It marks an extreme point in his music, where Shakespearean and literary ideas have no place; all is subsumed in a vision of humanity in collective obeisance to the presence of God.

The feeling for space in the *Te Deum* is expressed by the contrast of the organ with the orchestra and chorus. The organ should be at a distance from the rest and is not often heard simultaneously with them; the opening chords particularly exploit the directional idea. There are parts for two choruses and an extra body of 600 children's voices, in a manner similar to the ripieno line in the opening chorus of Bach's *St Matthew Passion*. Counterpoint again plays an important part in the formulation of the style. The 'Dignare' is constructed on a highly original device of moving the bass line through a succession of pedals, a 3rd apart, and Berlioz's technique of harmonic variation is much in evidence. The full forces produce moments of great dynamic impact, especially in the 'Tibi omnes' at the conclusion of each of three verses, and in the 'Judex crederis', described by Berlioz as 'Babylonian, Ninevitish', perhaps the most immense movement of his entire output: climax breaks over climax like an unending sea. The last movement is an orchestral march for the presentation of the colours, enacted at St Eustache in 1855, and an additional movement, never used by Berlioz, is a 'Prélude' designed for military occasions only. A tenor soloist sings in the 'Te ergo quaesumus'.

There are other choral works in which the same monumental style is applied on a narrower scale, for example the *Hymne à la France* (1844) and *L'impériale* (1854) whose titles betray their patriotic origins. The *Chant sacré* (1829) and the *Méditation religieuse* (1831), both settings of Thomas Moore, can be coupled as contemplative works, short but broad in style. *La révolution grecque* (1825–6) and *Le cinq mai* (1835) are more narrative, like dramatic cantatas. *Sara la baigneuse*, to a text by Victor Hugo, especially in its version for three separate choruses and small orchestra, is exquisitely poetic, one of Berlioz's most delicate and refined compositions. A number of choral works were composed with piano accompaniment, and the best of these are *Le ballet des ombres* (1828), a remarkably daring evocation of nocturnal spirits, and the *Chant guerrier* and the *Chanson à boire*, both in the *Irlande* collection of 1829, both exploiting expressive contrast as an element of form.

IV Songs

Some of Berlioz's vocal compositions, for example *Hélène* or *Sara la baigneuse*, exist in versions for four voices with accompaniment. There are songs for two or for three voices, so that the same phenomenon of a continuum between genres can be observed with choral music and songs. *La belle voyageuse* is a solo song with piano or orchestra, or a chorus for women's voices and

orchestra; *La mort d'Ophélie* is a solo song with piano
or for women's chorus with either piano or orchestral
accompaniment. Berlioz made adaptations according to
need wherever the expressive content of the piece allowed. A number of songs were orchestrated, and some,
like *Zaïde* and *Le chasseur danois*, came into being in
both piano and orchestral versions at the same time.

As a songwriter Berlioz owed much to the tradition
of the French *romance*, with which he was familiar from
childhood, and many of his earliest compositions were
in this mould. It is interesting to see how *La captive*,
composed in Italy in 1832, was originally a strophic
song relying on an exquisitely shaped melody, but was
later revised by Berlioz into a through-composed song
with orchestral accompaniment, a fully elaborated work
in his most expressive style. Even as late as 1850 he was
publishing songs like *Le matin* and *Petit oiseau* (two
settings of the same text) in an unambitious form with
simple piano accompaniment. One of his highest achievements in song is the *Elégie en prose*, the last of the
Irlande set of 1829, a fervent outpouring of Romantic
feeling. Berlioz wrote of it: 'I think I have rarely found a
melody of such truth and poignancy, steeped in such a
surge of sombre harmony'. But this is overshadowed by
the *Nuits d'été* of 1840–41, six settings of poems by
Gautier, originally composed for single voice with
piano, but orchestrated with some transpositions for
different voices in 1856. One should be wary of treating
the set as a strict cycle and Berlioz never performed it as
such, but it has a wholeness of mood and feeling and a
satisfying emotional balance. The outgoing mood of
Villanelle and *L'île inconnue* frame more sombre reflections on disappointed love, the longing of *Absence* and

the icy serenity of *Au cimetière*. The orchestral versions are executed with supreme skill, with light yet richly coloured textures throughout.

V Orchestral music

Besides the symphonies, Berlioz's orchestral output included five concert overtures that reflected Beethoven's and Mendelssohn's treatment of the overture as an independent form. *Waverley* (1828) and *Rob Roy* (1831) are based on Scott novels without any supposition that they preface an opera. In *Waverley* the contrast of slow introduction with vigorous Allegro is an illustration of the couplet:

> Dreams of love and Lady's charms
> Give place to honour and to arms.

It is one of the few works where Berlioz shows any affinity with the Italian style, but it is also experimental in feeling, especially at the beginning of the coda. *Rob Roy* was rejected by the composer as 'long and diffuse', which is a fair summary, and two themes were re-used in *Harold en Italie*. *Le roi Lear*, composed just before it, displays, in contrast, great concentration of energy; it is not a retelling of the play but a general representation of its mood, with events and characters worked into a logical musical sequence. Its composition was a spontaneous response to first reading the play and has some of the energetic turbulence of the first movement of the *Symphonie fantastique*. The overture to *Benvenuto Cellini* established a formal pattern of brief allegro before a slow section returning to the main allegro, and Berlioz used this in all his subsequent overtures, *Le carnaval romain*, *Le corsaire* and *Béatrice et Bénédict*.

Le carnaval romain is perhaps Berlioz's most extrovert and brilliant orchestral work, whose pace and glitter have long established it as a favourite concert showpiece. *Le corsaire* has a similar swiftness and brilliance, and a beautifully expressive slow section that is recalled, at fast tempo, in the Allegro. The music spells out the atmosphere and associations of the sea, in particular the Mediterranean, which provided Berlioz's first experience of wind and rigging in combat.

The *Marche funèbre pour la dernière scene d'Hamlet* is a neglected work, but one of his finest. There is a part for wordless chorus intoning a lugubrious 'Ah!' from time to time, but the burden of a long relentless climax is carried by an orchestral ostinato and a melody of hollow solemnity, the feeling so clearly and so often inspired in Berlioz by his experience of the play. The closing pages, where he used chromatic harmony to fine effect, are as affecting as Dido's final scene in *Les troyens* and seem to have been conceived on an equivalently broad scale.

Another solitary orchestral work is *Rêverie et caprice* for solo violin with piano or orchestral accompaniment; its restless movement, alternating rapidly between slow and fast tempos, is explained by its origin as a solo aria from *Benvenuto Cellini*, but it is hardly satisfactory as a violin showpiece and has too fragmentary a construction to be convincing.

VI Other works
Only with Berlioz's attitude to the mixing of opposites could a composition such as *Lélio* have come into existence, for its six musical numbers are, if regarded as separate entities, wholly diverse in subject and treatment. Yet the whole is given a sense of order by its

literary format and by the vivid links with Berlioz's personal life in each movement. Originally entitled *Le retour à la vie* in 1831, it was a pendant to the *Symphonie fantastique*, a further episode in an artist's life, showing how he comes to terms with life after an overwhelming traumatic experience, largely through the healing power of music and of creative fantasy. The *idée fixe* is also used to recall the symphony at crucial points. Much of the music of *Lélio* had existed before; for example, the 'Chant de bonheur' and 'La harpe éolienne' are revised from *La mort d'Orphée*, the Prix de Rome cantata of 1827. Yet the *Tempest* fantasy was incorporated unchanged. The relevance of each movement is made plain in the monologues between them, summarizing Berlioz's obsessions with Shakespeare, especially *Hamlet*, with brigands as a symbol of the free life, with passionate identification of self with others; these give it, quite apart from its unusually heterogeneous musical form, a unique place in the territory occupied by both literature and music.

Berlioz assembled other miscellanies with looser internal associations. There are the nine Moore settings of *Irlande* and the six Gautier settings of *Les nuits d'été*. Two *Hamlet* pieces and one Thomas Moore setting were grouped as *Tristia* for publication in 1850. Other groupings, largely for publisher's convenience, were *Feuillets d'album*, *Vox populi* and *Fleurs des landes*.

Three of the *Lélio* pieces were derived from Prix de Rome cantatas, of which Berlioz wrote four. The best of these are the first and third. *La mort d'Orphée*, of 1827, contains a bold Bacchanale and an affecting 'Tableau musical' at the end. The more conventional *Herminie*, of 1828, won second prize. *Cléopâtre*, of 1829, is start-

151

lingly dramatic with clear adumbrations of Berlioz's later tragic heroines, Juliet, Cassandra and Dido. The invocation, where she addresses the spirits of the Pharaohs, is magnificent. When re-used in *Lélio*, Berlioz's description of it was: 'Sombre orchestration, broad, sinister harmony, lugubrious melody ... a great voice breathing a menacing lament in the mysterious stillness of the night'. The fourth cantata, *La mort de Sardanapale*, is mostly lost, but was never highly valued by its composer.

Style

Berlioz's style of composition is recognized as one of the most idiosyncratic of the 19th century. It is quickly recognizable and has been as much reviled by his enemies as vaunted by his partisans. It is true that its characteristics do not always take immediate effect and that a familiarity with his music is often regarded as essential to its understanding. For a long time the inaccessibility of many of his scores produced a correspondingly hesitant public response, but the higher standing now enjoyed by Berlioz's music is reinforced and consolidated by its wider circulation. Opinions vary widely over the relative parts played in Berlioz's style by technique and inspiration. That he was subject to inspiration in truly Romantic fashion has never been denied, but this created the extraordinary belief that he had no technique and composed in a kind of blind fury. The truth is that Berlioz's music would be worthless with neither inspiration nor technique and that its mastery is due to an abundant provision of both. Both were of an unconventional kind; neither can be overlooked or denigrated for the advantage of the other. Certain elements of Berlioz's style call for separate discussion.

I Melody

Berlioz was a natural melodist. Few of his melodies fall into regular phrase lengths, and when they do, as in the second subject of *Les francs-juges* overture or in the *idée fixe* of *Harold en Italie*, they sound uncharacteristic. He found the regular balance of four- and eight-bar phrases uncongenial and spoke naturally in a kind of flexible musical prose, with surprise and contour important elements. His melodies sometimes expand to great length as at the opening of the *Symphonie funèbre et triomphale*, or fill out a whole musical movement in one long arch, as in Aeneas's 'Ah! quand viendra l'instant des suprêmes adieux', where internal repetition is minimal. The *idée fixe* of the *Symphonie fantastique* is well known for its expansive length. Much of Berlioz's melodic strength is built on small chromatic inflections, especially when an otherwise diatonic melody is slightly coloured by a chromatic note (with an attendant harmonic surprise). He was fond, for example, of falling chromatically from the 5th of the scale (G in the key of C), or falling chromatically downwards towards the 5th, so that using A♭, often in alternation with A♮, in C major or C minor is a recurrent fingerprint. The flattened 6th, especially in a major context, introduces a feeling of melancholy or loneliness, and a number of movements end with this almost unresolved hovering over the dominant note. The opening melody of *La damnation de Faust* offers a fine example of Berlioz's flattened 6th. There is an occasional modal touch in his melodies, especially in *L'enfance du Christ*, but he eschewed the folk idiom altogether. The sharpened 4th in the melody of *Le roi de Thulé* is a deliberately sophisticated attempt to portray Marguerite's naive nature, not the adoption of a naive style of his own.

II Harmony

Berlioz's understanding of harmony has been greatly

abused by those who have sought either a Brahmsian orderliness or a Lisztian spirit of adventure. By the standards of Chopin and Liszt the actual vocabulary of his harmony is restrained and there are few instances of enriched chromatic harmony. Berlioz was for the most part content with the harmonic vocabulary of Gluck and Beethoven, but he differed from most of his contemporaries in seeing harmony as an expressive rather than functional element. Chords do not lead one into another with inexorable cadential progress. They play their part one by one and become altered, when alteration is necessary, by the replacement of any or all of their notes. The element of surprise is intrinsic, for it is frequently the unexpected note of a chord that alters, and despite Berlioz's avowed dislike of enharmonic change he used it constantly. He similarly disliked accented appoggiaturas for creating new harmony, though many are to be found in his music. Diminished 7ths and kindred secondary 7th chords are much used, but generally without tonal pull. The suggestion that his harmonic thinking was derived from his study of the guitar has too weak a technical basis to be convincing.

A characteristic sonority is the grouping of upper parts as high as possible with the bass line isolated at a distance from them; at such times the strong melodic role of the bass line becomes evident. Much has been said about Berlioz's 'false' basses and his love of root positions, both of which are clear misrepresentations. A root position is sometimes disturbing when it anticipates a cadence on to the same root, but Berlioz preferred a smooth, often stepwise, movement to the striding pattern of a functional bass. The bass line is in free counterpoint with the upper line, with harmonic filling.

155

III Counterpoint

The free contrapuntal relationship of parts, especially the upper and lower parts of a texture, is one of the essential elements of Berlioz's style. In his mature compositions he exhibits plainly a distaste for 'tune with accompaniment', a mannerism associated by him with Italian opera and only used for special purposes, for example 'Un bal' in the *Symphonie fantastique* or Teresa's cavatina in *Benvenuto Cellini.* Contrapuntal textures are often an extension of orchestral textures, with layers seen in both contrapuntal and colouristic relationship to one another.

Berlioz regarded strict contrapuntal forms as mechanistic and inexpressive. He parodied the Handel–Cherubini style of fugue in *La damnation de Faust* and *Béatrice et Bénédict.* On the other hand fugato occurs repeatedly, generally to fine expressive and formal effect. There are choral fugues in both the Requiem and the *Te Deum,* whose 'Judex crederis' exhibits a type of fugue learnt from his teacher Reicha: the entries are successively one semitone higher. 'Châtiment effroyable' in Act 1 of *Les troyens* has the entries successively one tone lower. Berlioz's orchestral fugatos range widely, from the turbulent effect of the 'Ronde du sabbat' in the *Symphonie fantastique* and the middle section of *Le carnaval romain* to the wonderfully atmospheric fugal openings of the first and second parts of *La damnation de Faust* or the 'Chasse royale et orage' in *Les troyens,* whose fugal beginning is concealed in delicate harmony. Fugato is used to express strife in *Roméo et Juliette* and the streets of Jerusalem at night in *L'enfance du Christ,* and for a host of other purposes elsewhere. Canon and inversion are rare.

One type of contrapuntal treatment appealed greatly to Berlioz and this he called the 'réunion de deux thèmes' where two separate themes are heard first separately and then in combination. There are fine examples in the finale of the *Symphonie fantastique*, where the 'Ronde du sabbat' is combined with the *Dies irae*; in *Harold en Italie*; in the 'Fête chez Capulet' of *Roméo et Juliette*; in the overture to *Benvenuto Cellini*; in Act 4 of *Les troyens*, and elsewhere. The chorus of soldiers and students in *La damnation de Faust* is brilliantly effective, for the soldiers sing in B♭ major, in 6/8, and in French, while the students' song is in D minor, 2/4, and in Latin. At the opening of the carnival in *Benvenuto Cellini* Berlioz superimposed three separate elements, all distinct in character; this technique is a clear example of his belief in the combination of opposites and the mingling of diverse genres in a single work.

IV Rhythm

The vitality of Berlioz's music is to a large extent due to the clarity and boldness of his rhythmic articulation. In the mid-1830s, especially in *Harold en Italie* and *Benvenuto Cellini*, he exploited experimental rhythms, not just unusual time signatures but also superimpositions of different rhythms. Fieramosca's air in *Benvenuto Cellini* is an exercise in shifting time signatures. The 'Danse cabalistique' in *L'enfance du Christ* is in 7/4, the 'Combat de ceste' in *Les troyens* in 5/8. Berlioz did not succumb to the universal passion for triple metres, which his generation suffered, despite his recurrent fondness for 3/4 for music of tenderness or longing. He felt strongly that rhythm was inadequately studied by both composers and performers. As a conductor, too, he was noted for his rhythmic precision.

V Orchestration

Berlioz's long-standing reputation as a supreme orchestrator has sometimes overshadowed his other gifts. Instrumental colour is fundamental to his music; he was no pianist and never thought of sound, as Chopin and Brahms did, through the filter of the piano. But he played no orchestral instrument either (having abandoned the flute) and had to learn this art by studying textbooks, tutors, the instruments themselves, the scores of other composers, and by befriending players. Kástner's orchestration treatise of 1837 is the main significant predecessor to Berlioz's own, published in 1843 with a second edition in 1855. For Berlioz it was a sin to neglect the possibilities of orchestral instruments or to use them in unsuitable combinations. He was particularly anxious to use new instruments and took a close interest in Adolphe Sax's work. Instruments that had previously been used for special purposes he introduced into his normal requirements: the harp, for example, and the english horn are found in most of his scores; he was one of the first to write for the bass clarinet, the valve trumpet and the saxhorn; he made a special arrangement for the newly invented saxophone in 1844 and called for tuned cymbals in *Roméo et Juliette*; he required a piano, with two players, in the 'Tempest' fantasy in *Lélio*; there is a Turkish crescent in the *Grande symphonie funèbre et triomphale* and an antique sistrum in *Les troyens*.

But it is not the novelty of the instruments themselves that mark out Berlioz's orchestration so much as his skill in using them. Sometimes one instrument is used for a solo of striking fitness, for example the viola in *Harold en Italie* or the english horn in Marguerite's *romance* in *La damnation de Faust*. More often it is in combining and contrasting instruments that his judgment is most acute and inventive, especially in his use of wind. He wrote for woodwind in layers more often than in solos, and he liked the sound of wind chattering on repeated notes. Consider the 'Menuet des follets' in *La damnation de Faust* where the banks of woodwind give a splendidly rich effect followed by the darting brilliance of the two piccolos; in contrast there is the sombre colour of Romeo's arrival at the Capulets' vault, or of the 'Choeur d'ombres' in *Lélio*. Brass can be solemn or brazen; the 'Marche au supplice' in the *Symphonie fantastique* is a defiantly modern use of brass. Trombones introduce Mephistopheles with three flashing chords or support the gloomy doubts of Narbal in *Les troyens*. With a hiss of cymbals, *pianissimo*, they mark the entry of the Cardinal in *Benvenuto Cellini* and the blessing of little Astyanax by Priam in *Les troyens*.

There are innumerable instances of felicitous orchestral colour in Berlioz's music, and the delicacy of his use of *pianissimo* (as in the Queen Mab scherzo) is as memorable as the force of his immense sounds (as in the Requiem or the *Te Deum*). Yet he could also miscalculate, and there are occasions when the correct balance is extremely difficult to achieve, or when acoustics hinder the proper realization of a novel idea. In the latter category must be placed the trombone and flute chords in the Requiem and the timpani chords in the *Symphonie fantastique* and the Requiem, effective though they are from the expressive point of view. The influence of his orchestration has been immense, directly upon Liszt, Wagner, the Russians, Strauss and Mahler, but more profoundly by his emancipation of the procedure of orchestration. For Berlioz it was intrinsic to composition, not something applied to finished music. Berlioz also disregarded the 18th-century conception of orchestration as similar to part-writing for voices; in his hands timbre became something that could be used in free combinations as an artist might use his palette, without bowing to the demands of line, and this leads to the rich orchestral resource of Debussy and Ravel.

VI Space
A related element of Berlioz's style is his attention to the spatial distribution of sound. He believed strongly that music should be fitted to the building in which it is heard and he severely castigated the sound of noisy orchestras in small theatres. His scores, especially *Roméo et Juliette*, are filled with directions for the placing of players and singers. He was fond of offstage music, not only in the operas, but in the symphonies too: the shepherd's pipe is heard offstage in the *Symphonie fantastique*, the pilgrim's march is heard *au lointain* at the end of *Harold en Italie*; in *L'enfance du Christ* the angels are in a neighbouring room whose door is gradually closed. The Requiem is the grandest example of wide orchestral distribution and both the *Te Deum* and the *Symphonie funèbre et triomphale* exploit the distinct separation of parts of the orchestra. At the beginning of Act 1 of *Benvenuto Cellini* and Act 2 of *Béatrice et Bénédict* a great deal happens offstage, and the first-act finale of *Les troyens* is constructed on an elaborate panoply of three offstage groups carefully scored to suggest the approach and passing of the

Wooden Horse into the city. The offstage trumpets and drums in Marguerite's *romance* in *La damnation de Faust* simultaneously exploit their separateness in space and their total distinctness in musical language, diversity doubly expressed in both spatial and musical terms.

VII Form

In matters of form Berlioz paid only lip service to such inherited patterns as sonata form. Intuition and expression were allowed to dominate expectation and rule. Thematic development is abundant but irregular, tonal balance is felt rather than preordained. Large-scale tonal designs are not easy to discern; indeed there is no reason to expect them. Two movements in *Les troyens*, Andromache's scene in Act 1 and the Sentinels' scene in Act 5, do not end in the key in which they began, although they are musically self-contained; a similar case is the *Symphonie funèbre et triomphale*. The dominant does not necessarily play its classical role as antithesis to the tonic, and the notion of a 'second subject' in the symphonies is not always applicable. Development for its own sake, as a purely musical procedure, he avoided; he preferred to throw the weight of a movement on to the coda, or sometimes on to a series of codas of cumulative impact, and the sense of climax and closure is always strong. There are few musical forms as satisfying musically or emotionally as the 'Chasse royale et orage' (from *Les troyens*), yet apart from its return to C major, where it began, its tonal scheme is free and unfettered by pre-set schemes. Modulation in Berlioz's music is always fluid. Mediant relationships of every kind abound, so do Neapolitan and closer tonal shifts, made possible by his open attitude to the directional sense of harmony.

Two structural techniques should be mentioned. Berlioz contributed much to the then current desire to relate movements to each other by thematic, dramatic and other means. Thematic transformation is clearly seen in the *idée fixe* of the *Symphonie fantastique*, and in the treatment of many themes in *La damnation de Faust* and *Les troyens*. To change the significance and colour of a theme by adjusting its pace, pitch, metre or orchestration, was a technique Berlioz applied with great subtlety, as for example in the opening of Act 4 of *Les troyens* and the air of Narbal that follows. More personal to Berlioz is the device of harmonic variation, where a theme is presented against a series of different harmonies. The clearest example is the 'Tibi omnes' in the *Te Deum*, where the three strophes have the melody presented in three guises.

VIII Programmes

Berlioz's use of programmes throughout his music must be clearly understood as a natural outcome of his belief in the implicit kinship, identity even, of music and ideas. Since music was not autonomous, it must have equivalences and meanings in the world of action and imagination. In his mind music and literature were inextricably entwined, both expressions of the human soul. Poetry and literature often suggested music, music always suggested life and feeling.

It is thus absurd to speak, as many have, of Berlioz's 'reliance' on programmes, or of his use of them as propaganda. They are not there to serve the music as a means of making it more palatable or more intelligible.

They are part of it; they too reflect the movement of the composer's mind. He did not write the programme of the *Symphonie fantastique* in order to make it more sensational – that was scarcely needed; he wrote it because he felt it as part of the impulse that brought the music to birth. His programmes have the same status as his vocal texts. Many, of course, are not so explicit; indeed a title often serves as the sole direction, but the title or the image is always there (the *Toccata* for harmonium is the only exception). In some cases, as in the death scene in *Roméo et Juliette* or the 'Chasse royale et orage' in *Les troyens*, the action described by the music is continuous and precisely detailed.

A problem is presented by Berlioz's recurrent habit of self-borrowing, which generally arose from the desire to find better use for music first placed in an unsuitable, unfinished or unsuccessful setting. Generally there is no real conflict between the expressive purpose of one context and another, for it is clear that music of a given type can express many kinds of poetic or pictorial image and that therefore successive images may evoke or require the same music. The same image may likewise relate to more than one musical setting, although this is rarer. Self-borrowing is common because many of his ideas were unrealized, for a variety of reasons, and because he recognized the vitality of pieces that could otherwise be wasted. There is no evidence that when he was borrowing most heavily he was suffering any lack of fecundity, although the *Symphonie funèbre* came suspiciously close to a fallow period.

CHAPTER FIVE

The critic

Berlioz's views appeared regularly in the Paris press, and his literary output was huge. Apart from the *Traité d'instrumentation* and the *Voyage musical*, he published three collections of criticism: *Les soirées de l'orchestre* (1852), *Les grotesques de la musique* (1859) and *A travers chants* (1862). In his *feuilletons* he wrote of new operas and singers, many of them of staggering unimportance; his opinion on momentous occasions was of crucial interest, for example at the première of *Le prophète* in 1849. He reviewed most of the concerts of the Société des Concerts; he wrote of new instruments and musical gadgets, of his own impressions of music abroad, and of important musicians visiting France; he wrote biographical notices of Gluck, Beethoven, Spontini, Méhul and himself; he wrote fiction and fantasy, often with a critical purpose; he wrote serialized treatises on orchestration and conducting. There are, in short, few facets of musical practice of the time untouched in his *feuilletons*.

Inconsistencies and changes of opinion are to be found, as one would expect over 30 years; but in general Berlioz's opinions are trenchant and clearly expressed. He loathed the easy success of second-rate musicians with no personality and a borrowed style, and he fought endlessly against backstage politics that placed graft above art. His admiration for the greatest masters, especially Gluck and Beethoven, is a leitmotif of almost wearying persistence, and with secondary masters, such as Rossini, Meyerbeer and Halévy, he was carefully discriminating, separating the good from the bad. He greeted Glinka and Bizet with prophetic enthusiasm, yet Wagner ultimately taxed his deep-rooted beliefs beyond the boundaries of acceptance. The *Tristan* prelude had, for him, 'no other theme than a sort of chromatic sigh'. It was 'full of dissonant chords, the harshness of which is intensified by modifications of the real notes of the harmony'; he acknowledged Wagner as a powerful new voice, but one that was speaking a language he no longer understood, and leading the next generation away from the highest reaches of the art.

Berlioz was one of the first to enunciate a critical standpoint that is now a commonplace but was then startlingly new: that music should be enshrined in the form in which it was written and not brought up to date. He attacked Fétis and Habeneck for their 'corrections' of Beethoven, and repudiated singers who added ornamentation and 'improvements' to the vocal lines of Gluck and Mozart. For Berlioz the composer's utterance had a sanctity that raised it above the tampering of mere performers. That is not to say that he only accepted whole performances, for his concert programmes were full of extracts, as was the custom of the day, but the principle of respecting a composer's own directions had his constant support. Castil-Blaze's travesties of Mozart and Weber appalled him, and it was an ironic twist that exposed him to the criticism of having mutilated Weber's *Der Freischütz* when he had set the recitatives to music in order to prevent the Opéra from mutilating it any more. A series of sarcastic directions in the autograph of *Les troyens* permits cuts to be made when circumstances render them necessary; a footnote in *Roméo et Juliette* advises suppression of the more demanding sections when the audience is not sufficiently attuned to the composer's purpose.

Berlioz's battles as a critic were not fought just to expound his points of view. They were intimately related to the more serious struggle for recognition as a composer. If his readers could be persuaded to recognize the good and the beautiful in Gluck, Spontini and any modern composer, so they might turn more sympathetically to his own music. But the strategy failed and his journalism was seen more and more as an independent professional activity, executed with extreme flair and wit, but in fact making his stature as a composer all the harder to establish.

CHAPTER SIX

Influences and research

Berlioz belongs to a tradition, yet nevertheless he is an isolated figure. Since the music on which he based his style is now little known, he has been regarded too simply as a wholly unprecedented phenomenon in French music. From his predecessors he inherited a basic language and certain mannerisms, for example dramatic recitative gestures from Spontini and a taste for plain melody from Gluck. Méhul's raw vigour is to be seen echoed in Berlioz's early music, and Le Sueur's passionate search for new modes of expression, by using lengthy descriptions or unusual instrumental effects, left a clear mark. The grandiose music of the French Revolution, especially such pieces as the *Marche lugubre* of Gossec, is carried on in Berlioz's monumental style. He learnt much from Weber and Beethoven and a little, despite himself, from Rossini; his contemporaries on the whole did not influence him greatly. The shape and pulse of his themes and their treatment, his sense of colour and contrast, the urgent flux of passion and the immense expressive variety of his music: these were all new. No other French composer of his time had the imagination or the genius to grasp the Berliozian manner, which was in any case too personal to permit easy imitation. German composers, like Schumann, admired his music but spoke a different language. Mendelssohn admired him as a man, but disliked his music.

167

Berlioz's influence was most obviously shown (to the point of imitation) by such minor figures as David and Reyer in France and Cornelius in Germany. The latter's *Barbier von Bagdad* is full of homage to *Benvenuto Cellini*. More important was the fertilization of Liszt's music, shown especially in the symphonic poems, a debt Liszt was glad to acknowledge. Wagner stands clearly in line, yet though he adapted a number of felicitous inventions from Berlioz and can be shown to have learnt much from him, his fundamental outlook was too different and too all-embracing to be regarded as an offshoot of Berlioz's Romanticism in particular.

The Russians adopted Berliozian ideas with enthusiasm, especially Balakirev, whose plan for a *Manfred* symphony, intentionally modelled on *Harold en Italie*, was taken up by Tchaikovsky. Rimsky-Korsakov's *Antar* and many other poematic symphonies show a debt to Berlioz. Strauss showed it, too, in *Aus Italien* and especially *Don Quixote*. In France Berlioz's style effectively had no influence on the succeeding generations. Both Debussy and Ravel repudiated him on technical grounds, although more sympathetic attitudes have been voiced by Milhaud and Messiaen. In sum, it is a sorry tale of rejection and isolation. Berlioz has inspired many by the sincerity and energy of his music, but in his lifetime the opportunity of absorbing even part of an idiosyncratic style was missed. As an idealist he had much to offer to artists of any milieu, and history has forced one to recognize him for what he was and what he did rather than for where he stood in relation to others. This may ultimately prove a blessing.

Much literature was published on Berlioz during his lifetime, but the first full-length biographies were written by Jullien and Hippeau in the 1880s. The centenary of 1903 coincided with a wave of Berliozian study that produced the Breitkopf & Härtel collected edition, Adolphe Boschot's three-volume biography and a wide range of special studies by Prod'homme and Tiersot, who also published three volumes of correspondence. The greatest resurgence of interest has been later in the century, assisted particularly by recordings, by Barzun's two-volume study of 1950, and by the London revival of *Les troyens* in 1957. By the 1969 centenary a new understanding was at last possible from a wider familiarity with the music, especially *Les troyens*, which was finally performed, published and recorded in full for the first time. New complete editions of the music, the literary works and the correspondence are all now in progress, so that a full and fair presentation of Berlioz's life and work may eventually atone for many years of neglect and misunderstanding.

WORKS

Editions: H. Berlioz: Werke, ed. C. Malherbe and F. Weingartner (Leipzig and elsewhere, 1899–1907/R) [B&H]
New Berlioz Edition, ed. H. Macdonald and others (Kassel, 1967–) [NBE]
Catalogue: Catalogue of the Works of Hector Berlioz, ed. D. K. Holoman (in preparation) [H]

Numbers in the right-hand column denote references in the text.

OPERAS

op.	Title and genre	Acts and libretto	Composed	Published	First performance	Remarks	B&H	NBE	H	134-8
[3]	Les francs-juges	3, H. Ferrand	1826	ov., op.3, 1836	not perf.	rev. 1829; rev. 1833 as Le cri de guerre de Brisgau (I, T, Gounet); 5 complete movts extant	ov., iv	iv	23	88, 92, 106, 109, 124, 134, 135, 137, 141, 153
—	Benvenuto Cellini, opera semiseria	2, L. de Wailly and A. Barbier	1834–7	vocal score, 1856	Paris, Opéra, 10 Sept 1838	rev. 1852 (3 acts), Weimar, 17 Nov 1852	ov., v	i	76	100, 106, 109, 112, 117–18, 122, 135, 137, 140, 149, 150, 156, 157, 159, 160, 168
—	La nonne sanglante	Scribe	1841–7		not perf.	not completed	—	iv	91	109, 137
—	Les troyens, grand opera	5, Berlioz, after Virgil	1856–8	vocal score, 1861	Paris, Théâtre-Lyrique, 4 Nov 1863 (II); Karlsruhe, 6 Dec 1890 (I)	enlarged and rev. 1859–60; divided into (I) La prise de Troie, (II) Les troyens à Carthage, 1863, the latter with prol	—	ii	133	95, 96, 100, 118, 122–3, 124, 136, 137–8, 145, 150, 156, 157, 158, 159, 160, 161, 162, 163, 166, 169
—	Béatrice et Bénédict, opera	2, Berlioz, after Shakespeare	1860–62	vocal score, 1863	Baden-Baden, 9 Aug 1862		xix–xx	iii	138	89, 95, 106, 119, 126, 137, 149, 156, 160

SYMPHONIES

op.	Title and genre	Text	Forces	Composed	Published	Remarks	B&H	NBE	H	
14	Symphonie fantastique: épisode de la vie d'un artiste		orch	1830	1845		i	xvi	48	89, 92, 93, 95, 96, 97, 98, 101, 102, 103, 105, 112, 129, 134, 135, 139–40, 142, 149, 151, 154, 156, 157, 159, 160, 162, 163
16	Harold en Italie, symphonie en 4 parties		va, orch	1834	1848		ii	xvii	68	100, 105, 107, 126, 140, 149, 154, 157, 159, 160, 168
17	Roméo et Juliette, symphonie dramatique	E. Deschamps, after Shakespeare	A,T,B, STBSTB, orch	1839	1847		iii	xviii	79	95, 96, 100, 107, 113, 114, 115, 118, 140–41, 143, 156, 157, 158, 159, 160, 163, 166
15	Grande symphonie funèbre et triomphale	A. Deschamps	military band, str and SSTTBB ad lib	1840	1843	last movt arr. Mez/T, vv, pf, 1848, as L'apothéose	i xvi	xix xiv	80	104, 107, 112, 141, 143, 154, 158, 160, 161, 163

OTHER ORCHESTRAL

op.	Title and genre	Composed	Published	Remarks	B&H	NBE	H	
1	Waverley, grande ouverture	by May 1828	1839		iv	xx	26	92, 97, 149
4	Le roi Lear, grande ouverture	1831	1840		iv	xx	53	95, 102, 149
—	Intrata di Rob Roy Macgregor	1831	1900		iv	xx	54	102, 149
8	Rêverie et caprice, romance	1841	1841	for vn, orch/pf; arr. from cavatina composed for Benvenuto Cellini	vi	xxi	88	150
9	Le carnaval romain, ouverture caractéristique	1844	1844	on material from Benvenuto Cellini	v	xx	95	112, 149, 150, 156
21	Le corsaire, ouverture	1844	1852	orig. title La tour de Nice; 2nd title Le corsaire rouge; rev. before 1852	v	xx	101	112, 149, 150
—	Marche troyenne	1864	1865	arr. from Act 1 of Les troyens	vi	xxi	133B	

171

CHORAL WORKS

op.	Title and genre	Text	Forces	Composed	Published	Remarks	B&H	NBE	H.	143–7
—	Resurrexit		STTB, orch	1824	1902	only surviving part of the Mass of 1824; rev. 1829, as Le jugement dernier; largely absorbed into Benvenuto Cellini, Grande messe des morts and Te Deum	vii	xii	20	91
—	La révolution grecque, scène héroïque	H. Ferrand	B, B, SSATTBB, orch	1825–6	1903	2 sections rev. for vv, wind band, 1833	x	xii	21	92, 147
—	La mort d'Orphée, monologue et bacchanale	Berton	T, SSSS, orch	1827	1930	Prix de Rome cantata	—	vi	25	92, 151
2	Le ballet des ombres, ronde nocturne	A. Duboys, after Herder	STTBB, pf	1828	1829	withdrawn by Berlioz	xvi	xiv	37	147
1	Huit scènes de Faust	G. de Nerval, after Goethe		1828–9	1829	withdrawn by Berlioz; later used in La damnation de Faust	x	v	33	96–7, 113, 133, 143
	1 Chants de la fête de Pâques		SSSSTTBB, orch							
	2 Paysans sous les tilleuls, danse et chant		STTB, orch							
	3 Concert de sylphes		S, S, A, T, Bar, B, orch							97
	4 Écot de joyeux compagnons, histoire d'un rat		B,TTBB, orch							
	5 Chanson de Méphistophélès, histoire d'une puce		T,TTBB, orch							
	6 Le roi de Thulé, chanson gothique		S, orch			also for 1v, pf				154
	7 Romance de Marguerite, chœur de soldats		S, TTBB, orch							
	8 Sérénade de Méphistophélès		T, gui							97

2/3	Chant guerrier	T. Gounet after Moore	T, TBB, pf	1829	1830	9 mélodies (Irlande), no.3	xvi	xiv	41	97, 147, 151
2/5	Chanson à boire	Gounet, after Moore	T, TTBB, pf	1829	1830	9 mélodies (Irlande), no.5	xvi	xiv	43	97, 147, 151
2/6	Chant sacré	Gounet, after Moore	T, SSTTBB, pf/orch	1829	1830	9 mélodies (Irlande), no.6; 2 versions; orchd 1843	xiv, xvi	xii, xiv	44	97, 147, 151
—	La mort de Sardanapale	Gail	S, TTBB, orch	1830	—	Prix de Rome cantata, mostly lost	—	vi	50	93, 98, 152
18/1	Méditation religieuse	Berlioz, after L. Swanton Belloc's trans. of Moore	SATTBB, orch	1831	1852	Tristia, no.1; orig. acc. 7 wind insts [lost]	xiv	xii	56	147, 151
14bis	Lélio, ou Le retour à la vie, monodrame lyrique	Berlioz (except no.1)		1831–2	1855	orig. title Le retour à la vie, mélologue en six parties; sequel to Symphonie fantastique; rev. 1854	xiii	vii	55	102, 103, 133, 150, 151, 152, 158, 159
1 Le pêcheur (ballade)		A. Duboys, after Goethe	T, pf			adapted from song of ?1828				
2 Choeur des ombres			STB, orch			adapted from section of Cléopâtre				151
3 Chanson de brigands			B, TTBB, orch			adapted from lost Chanson de pirates, 1829				
4 Chant de bonheur			T, orch			adapted from La mort d'Orphée; also arr. T, pf				151
5 La harpe éolienne, souvenirs			orch			adapted from La mort d'Orphée				151
6 Fantaisie sur la Tempête de Shakespeare			SSATT, orch	1830	1830	perf. 1830 as Ouverture pour la Tempête de Shakespeare		vii	52	95, 98, 151, 158
Quartetto e coro dei maggi			SSTB, orch	1832	1902	possibly rev. of lost Marche religieuse des mages, 1828		xii	59	
11	Sara la baigneuse, ballade	Hugo	STBB, SA, TTBB, orch	1834	1851	rev. 1850 from lost orig. for TTBB/STTB, orch; also pubd for 2 vv, pf (1850)	xiv	xii	69	107, 147
13/5	Le chant des Bretons	A. Brizeux	TTBB, pf	1835	c1835	rev. as Fleurs des landes, no.5 (1850); both versions also arr. T, pf; an arr. for vv, orch is of doubtful authenticity	xvi	xiv	71	151

173

op.	Title and genre	Text	Forces	Composed	Published	Remarks	B&H	NBE	H	
6	Le cinq mai, chant sur la mort de l'empereur Napoléon	P. J. de Béranger	B, SSTTBB, orch	1835	c1840	refrain composed 1832	xiii	xii	74	107, 147
5	Grande messe des morts (Requiem)		T, SSTTBB, orch	1837	1838	rev. 1852, 1867	vii	ix	75	104, 105, 107, 109, 119, 145-6, 156, 159, 160
—	Choeur de 402 voix en langue celtique inconnue	Berlioz	SATB	1843	1969	albumleaf	—	xiv	93	
20/2	Hymne à la France	A. Barbier	SSAATTBB, orch	1844	1850	Vox populi, no.2	xiv	xii	97	112, 147, 151
2/2	Hélène, ballade	Gounet, after Moore	TTBB, orch	1844	1903	arr. of song orig. for 2vv	xiv	xii	40B	147
18/3	Marche funèbre pour la dernière scène d'Hamlet		SATB, orch	1844	1852	Tristia, no.3	vi	xii	103	95, 96, 150, 151
24	La damnation de Faust, légende dramatique	G. de Nerval, A. Gandonnière and Berlioz, after Goethe	Mez, T, Bar, B, SSATTBB, orch	1845-6	1854	incorporating rev. versions of Huit scènes de Faust	xi-xii	viii	111	54, 113, 114, 117, 118, 126, 143-4, 154, 156, 157, 159, 161, 162
19/3	Le chant des chemins de fer	J. Janin	T, SSTTBB, orch	1846	1850	Feuillets d'album, no.3	xiv	xii	110	114
20/1	La menace des Francs, marche et choeur		T, T, B, B, SSTTBB, orch	1848 or earlier	1850	Vox populi, no.1	xiv, xvi	xii	117	151
18/2	La mort d'Ophélie, ballade	E. Legouvé, after Shakespeare	SA, pf/orch	1848	1852	arr. from solo song; Tristia, no.2	xiv, xvi	xii, xiv	92B	95, 97, 148, 151
19/4	Prière du matin	Lamartine	SS [children], pf	1846 or earlier	1848	Feuillets d'album, no.4	xvi	xiv	112	151
22	Te Deum		T, STB, STB, S [children], orch	1849	1855		viii	x	118	120, 121, 145, 146-7, 156, 159, 160, 162
25	L'enfance du Christ, trilogie sacrée	Berlioz	S, T, T, Bar, B, B, B, SATB, orch	1850-54	1855		ix	xi	130	119-20, 121, 143, 144-5, 154, 156, 157, 160
	1 Le songe d'Hérode			1854						
	2 La fuite en Egypte			1850	1852					119, 120
	3 L'arrivée à Saïs			1853-4						
2/4	La belle voyageuse, ballade	T. Gounet, after Moore	SA, orch	1851	—	arr. of solo song	—	xiii	42D	147
26	L'impériale, cantate	Lafont	SATB, SATB, orch	1854	1856		xiii	xii	129	124, 147
—	Hymne pour la consécration du nouveau tabernacle	J. H. Vries	SSATTBB, pf/org	1859	1859		xvi	xiv	135	
28	Le temple universel	J. F. Vaudin	TTBB, TTBB, org	1861	1861	rev. TTBB, unacc., 1867	xvi	xiv	137	

op.	Title and genre	Text	Voice	Composed	Orchestrated	Published	Remarks	B&H	NBE	H	
—	Veni creator, motet		S, S, A, SSA, org ad lib			between 1877 and 1888		vii	xiv	141	
—	Tantum ergo		S, S, A, SSA, org			between 1877 and 1888		vii	xiv	142	

SOLO VOICE AND ORCHESTRA

op.	Title and genre	Text	Voice	Composed	Orchestrated	Published	Remarks	B&H	NBE	H	
—	Herminie, scène lyrique	P. A. Vieillard	S	1828		1903	Prix de Rome cantata	xv	vi	29	93, 151
—	La mort de Cléopâtre, scène lyrique	Vieillard	S	1829		1903	Prix de Rome cantata	xv	vi	36	93, 151–2
2/4	La belle voyageuse, légende irlandaise	T. Gounet, after Moore	Mez	1829	1842	c1844	orig. for Mez, pf	xv	xiii	42C	97, 147
12	La captive, orientale	Hugo	Mez/A	1832	1848	c1849	orig. for S, pf, earlier orchestration, 1834, lost; this version exists in 2 keys	xv	xiii	60E–F	103, 107, 148
13/4	Le jeune pâtre breton	A. Brizeux	Mez/T	1833		1839	orig. version, lost, used in Le cri de guerre de Brisgau, 1833; rev. 1835	xv	xiii	65D	107
—	Aubade	Musset	S/T	1839	?	1975	arr. for 1v, 2 cornets, 4 hn of song for 1v, 2 hn, 1839	—	xiii	78	
7	Les nuits d'été	T. Gautier					orig. for Mez/T, pf	xv		82–7	97, 107, 111, 122, 148–9, 151
	1 Villanelle		Mez/T	1840–41	1856	1856					148
	2 Le spectre de la rose		A	1840	1855 or 1856	1856					122
	3 Sur les lagunes		Mez/A/Bar	1840–41	1856	1856					111, 122, 148
	4 Absence		Mez/T	1840	1843	1843					149
	5 Au cimetière (clair de lune)		T	1840–41	1856	1856					
	6 L'île inconnue		Mez/T	1840–41	1856	1856					148
19/6	Le chasseur danois	A. de Leuven	B	1845	1845	1903	also for B, pf	xv	xiii	104B	113, 148
19/1	Zaïde, boléro	R. de Beauvoir	S	1845	1845	1903	also for S, pf [2 versions]	xv	xiii	108B	113, 148

175

SONGS

op.	Title and genre	Text	Forces	Composed	Published	Remarks	B&H	NBE	H	
—	Le dépit de la bergère, romance	Mme***	1v, pf	1819 or earlier	?1819		xvii	xv	7	89
—	Pleure, pauvre Colette, romance	Bourgerie	S, S/T, T, pf	?1818–22	1822		xvi	xv	11	
—	Canon libre à la quinte	Bourgerie	A, Bar, pf	?1818–22	1822		xvi	xv	14	
—	Le maure jaloux, romance	Florian	T, pf	?1818–22	1822	also entitled L'arabe jaloux	xvii	xv	9	
—	Amitié, reprends ton empire, romance et invocation	Florian	S, S, B, pf	?1818–22	1823		xvi	xv	10	
—	Toi qui l'aimas, verse des pleurs, romance	A. Duboys	T, pf	?1822–3	1823		xvii	xv	16	
—	Le montagnard exilé, chant élégiaque	Duboys	S, S, pf/harp	?1822–3	1823		xvi	xv	15	
—	Nocturne	Duboys	S, S, gui	?1825–30	—		—	xv	31	
1/6	Le roi de Thulé, chanson gothique	G. de Nerval, after Goethe	S, pf	1828	—	used in 8 scènes de Faust and La damnation de Faust	—	xv	33B	154
14bis	Le pêcheur, ballade	A. Duboys, after Goethe	T, pf	?1828	1833	used in Le retour à la vie (Lélio)	xvii	xv	55	
2/1	Le coucher du soleil, rêverie	T. Gounet, after Moore	T, pf	1829	1830	9 mélodies (Irlande), no.1	xvii	xv	39	97, 151
2/2	Hélène, ballade	Gounet, after Moore	S, S/T, B, pf	1829	1830	9 mélodies (Irlande), no.2; later arr. TTBB, orch	xvi	xv	40	97, 147, 151
2/4	La belle voyageuse, ballade	Gounet, after Moore	Mez, pf	1829	1830	9 mélodies (Irlande), no.4; later arr. TTBB, orch [lost], Mez, orch; later arr. SA, orch	xvii	xv	42A	97, 147, 151
2/7	L'origine de la harpe, ballade	Gounet, after Moore	S/T, pf	1829	1830	9 mélodies (Irlande), no.7	xvii	xv	45	97, 151
2/8	Adieu Bessy, romance anglaise et française	Gounet, after Moore	T, pf	1829	1830	9 mélodies (Irlande), no.8; 2 versions	xvii	xv	46	97, 151
2/9	Elégie en prose	Louise Belloc, after Moore	T, pf	1829	1830	9 mélodies (Irlande), no.9	xvii	xv	47	97, 148, 151
14bis	Chant de bonheur	Berlioz	T, pf	1831–2	1833	arr. from Le retour à la vie (Lélio); orig. from La mort d'Orphée	xvii	xv	55	
12	La captive, orientale	Hugo	Mez, pf	1832	1904	1st version	xvii	xv	60A–B	103, 107, 148
12	La captive, orientale	Hugo	Mez, vc, pf	1832	1832	2nd version	xvii	xv	60C	103, 107, 148

13/4	Le jeune pâtre breton	A. Brizeux	Mez/T, pf	1833	rev. with hn ad lib, 1935; also orchd	1835	xvii	xv	65A, C	107
19/2	Les champs, romance	P. J. de Béranger	T, pf	1834 or earlier	rev. in Feuillets d'album (1850)	1834	xvii	xv	67	151
—	Je crois en vous, romance	L. Guérin	T, pf	1834 or earlier	used in Benvenuto Cellini	1834	xvii	xv	70	
13/5	Le chant des bretons	Brizeux	T, pf	1834	rev. as Fleurs des landes, no.5 (1850); both versions also arr. TTBB, pf	c1835	xvii	xv	71	151
—	Chansonette	L. de Wailly	S/T, pf	1835	used for the Choeur de masques in Benvenuto Cellini	1974	—	xv	73	
—	Aubade	Musset	1v, 2 hn	1839	later rev. acc. 2 cornets, 4 hn	1975	—	xv, xiii	78	148
7	Les nuits d'été	Gautier	Mez/T, pf	1840–41		1841	xvii	xv	82–7	97, 107, 111, 122, 148–9, 151
	1 Villanelle				orchd 1856					148
	2 Le spectre de la rose				rev., orchd 1855 or 1856					122
	3 Sur les lagunes, lamento				orchd 1856					111, 122, 148
	4 Absence				orchd 1843					149
	5 Au cimetière, clair de lune				rev., orchd 1856					
	6 L'île inconnue				orchd 1856					148
18/2	La mort d'Ophélie	E. Legouvé, after Shakespeare	S/T, pf	1842	later arr. SA, orch/pf, pubd as Tristia, no.2	1848	xvii	xv	92A	95, 148, 151
19/5	La belle Isabeau, conte pendant l'orage	Dumas	Mez, pf	1843 or earlier	2nd version, Mez, SSTTBB, pf, 1844, pubd as Feuillets d'album, no.5 (1850)	1843	xvii	xv	94	151
19/6	Le chasseur danois	A. de Leuven	B, pf	1845	orchd 1845; Feuillets d'album, no.6	1845	xvii	xv	104A	113, 148, 151
19/1	Zaïde, boléro	R. de Beauvoir	S, pf, castanets ad lib	1845	2 versions: orchd 1845	1845	xvii	xv	108A	113, 148
13/3	Le trébuchet	A. Bertin and E. Deschamps	S, S/T, Bar, pf	1846 or earlier	Fleurs des landes, no.3	1850	xvi	xv	113	151
—	Nessun maggior, page d'album	Berlioz, after Dante	S/T, pf	1847		1904	xvii	xv	114	
13/1	Le matin, romance	A. de Bouclon	Mez/T, pf	1849 or earlier	Fleurs des landes, no.1	1850	xvii	xv	124	148, 151
13/2	Petit oiseau, chanson de paysan	Bouclon	T/Bar/Mez, pf	1849 or earlier	Fleurs des landes, no.2; words the same as for Le matin	1850	xvii	xv	125	148, 151

MISCELLANEOUS WORKS

op.	Title and genre	Text	Forces	Composed	Published	Remarks	B&H	NBE	H	
—	Fugue à 4 voix	—	4 pts.	1826	—	Prix de Rome submission	—	vi	22	
—	Fugue à 3 sujets	—	4 pts.	1829	1902	Prix de Rome submission	vi	vi	35	
—	Dans l'alcôve sombre	Hugo	?1v, pf	c1832	—	sketch	—	xxiii	62	137
—	Erigone, intermède antique	after Ballanche	solo vv, vv, orch	?1836–41	—	frags. only	—	xxiii	77	
8	Rêverie et caprice	—	vn, pf	1841	1841	also for vn, orch, arr. from cavatina composed for Benvenuto Cellini	vi	xxi	88	150
—	Andante in B	—	2 pts.	1842	—	albumleaf	—	xxiii	93	
—	Chasse à la grosse bête	—	ob, bn	1843	—	albumleaf	—	xxiii	93	
—	Feuillet d'album	—	pf	1844	—		—	xxiii	96	
—	Trois morceaux 1 Sérénade agreste à la Madone sur le thème des pifferari romains 2 Toccata 3 Hymne pour l'élévation	—	orgue-mélodium	1844	1844		vi	xxi	98	133
—	Le vent gémit, sérénade	Méry	1v	1845	—	2 versions	—	xxiii	99 100	162
—	Valse chantée par le vent dans les cheminées d'un de mes châteaux en Espagne	—	—	1855	—	albumleaf	—	xxiii	107 131	
—	Au bord d'une rivière	?	1v, pf			sketch	—	xxiii	132	
—	Salut matinal (en langue kanaque)	Berlioz	1v		1954	albumleaf	—	xxiii	140	

ARRANGEMENTS

Composer or source	Title		Forces and remarks	Arranged	Published	B&H	NBE	H	
Various			gui acc. for romances by Lintan, V. Martini, Dalayrac, Pollet, Catrufo, Bédart, Boieldieu, Della Maria, Plantade, Berton, Solié, Nadermann, Lélu, Messonier	?1818–21	—	—	xxii	8	106

Composer	Title	Forces	Date	Date			H	Bibl.
Rouget de Lisle	Hymne des Marseillais	1 TTB, SSTB, orch	1830	1830	xviii	xxii	51A	
Rouget de Lisle		2 T, SSTTBB, pf	1848	1848	—	xxii	51B	
F. Huber	Chant du neuf Thermidor	T, SSTTBB, orch	1830	1984	—	xxii	51bis	
	Sur les alpes quel délice (le chasseur de chamois)	3 male vv	1833	—	—	xxii	64	
Weber	Der Freischütz	dialogue composed to recit	1841	1842	xviii	xxii	89	109
Weber	L'invitation à la valse	orch	1841	1842	xviii	xxii	90	112
L. de Meyer	Marche marocaine	orch	1845	1846	—	xxii	105	113, 143
[trad.]	Marche de Rákóczi	orch; used in La damnation de Faust	1846	1854	xi	viii	109	
Bortnyansky	Adoremus (le chant des chérubins)	SATB; Lat. words by Berlioz	1850 or earlier	1851	xviii	xxii	122	
Bortnyansky	Pater noster	SATB; Lat. words by Berlioz	1850 or earlier	1851	xviii	xxii	126	
Martini	Plaisir d'amour	Bar, orch	1859	1859	xviii	xxii	134	
Schubert	Der Erlkönig	T, orch	1860	1860	xviii	xxii	136	
Couperin	Invitation à louer Dieu	arr. from Soeur Monique for SSA, pf	—	between 1877 and 1888	xviii	xxii	146	

LOST WORKS

Title and genre	Text	Forces	Composed	Remarks		H
Potpourri concertant sur des thèmes italiens	—	fl, hn, str qt	c1818		88, 133	1
2 qnts		fl, str qt	1823	1 melody used in ov. to Les francs-juges	88, 133	2, 3
Estelle et Némorin, songs	Florian		1822-3	1 melody used in Symphonie fantastique	89	6
Le cheval arabe, cantata	Millevoye	1v, pf	1822-3		91	12
Canon à trois voix		1v, orch	1823			13
Estelle et Némorin, opera	Gerono, after Florian		1823	probably using the earlier Florian songs	91, 134	17
Le passage de la mer rouge, Lat. oratorio	?		1823		91	18
Beverley ou Le joueur, scena	Saurin	B, orch	1824			19
Messe solennelle		Bar, vv, orch	1824	in 8 or 9 movts; Resurrexit survives	107	20
Les francs-juges, opera	Ferrand		1826	ov., 5 complete movts survive; the rest was destroyed	88, 92, 106, 109, 134, 135, 137, 141, 153	23
Fugues	—		1827-8	Prix de Rome submissions		24, 28
Marche religieuse des mages	—		1828 or earlier	Possibly related to Quartetto e coro dei maggi of 1832		27

Title and genre	Text	Forces	Composed	Remarks	H	
Variations on Mozart's Là ci darem la mano	—	gui	1828 or earlier	pubd by Aulagnier	30	
O salutaris		3 solo vv, org/pf	1828–9	probably the same as lost oratorio written for Choron	32	
Chanson de pirates	Hugo	?1v, orch	1829		34	
Fugue	?		1830	rev. as Chanson de brigands in Lélio	49	
Choeur d'anges pour les fêtes de Noël			1831	Prix de Rome submission	58	
Choeur de toutes les voix	Berlioz		1831		57	
Romance de Marie Tudor	Hugo	T, ?orch	1833	perf. 22 Dec 1833	66	
Fête musicale funèbre à la mémoire des hommes illustres de la France	?	?vv, orch	1835	2 movts of 7 completed; these probably incorporated the Resurrexit and Le dernier jour du monde (planned 1831–2) and were used in, probably, Le cinq mai, Benvenuto Cellini, the Grande messe des morts and the Grande symphonie funèbre et triomphale	72	107
Plain-chants de l'église grecque	—	16vv	1843	commissioned by the Russian imperial chapel	—	
[Hymne]	—	6 Sax insts	1844	probably arr. of Chant sacré	44C	
[Marche d'Isly]	—	orch	1845	arr. of Léopold de Meyer's Marche d'Isly listed as unpubd in Labitte catalogue, 1846;	108	
Ouverture des ciseleurs		orch	1846	probably based on Benvenuto Cellini; perhaps never written		

WRITINGS

(see Hopkinson, 1951, for more detailed bibliographical information)

Grand traité d'instrumentation et d'orchestration modernes, op.10 (Paris, 1843, 2/1855; Eng. trans., 1856/R1970) — 164–6

Voyage musical en Allemagne et en Italie (Paris, 1844/R1970) — 110, 129, 158, 164

Les soirées de l'orchestre (Paris, 1852; Eng. trans., 1956, 2/1973); ed. L. Guichard (Paris, 1968) — 110, 164

Le chef d'orchestre: théorie de son art (Paris, 1856; Eng. trans., 1917) — 30, 129, 164

Les grotesques de la musique (Paris, 1859); ed. L. Guichard (Paris, 1969) — 114, 164

A travers chants (Paris, 1862; Eng. trans., 1913–18); ed. L. Guichard (Paris, 1971) — 164

Mémoires de Hector Berlioz (Paris, 1870; ed. and Eng. trans. by D. Cairns, 1969, 2/1970); ed. P. Citron (Paris, 1969) — 88, 100, 101, 110, 112, 113, 116, 120, 124, 143

Les musiciens et la musique, ed. A. Hallays (Paris, 1903)

Cauchemars et passions, ed. G. Condé (Paris, 1981)

Articles in Le corsaire (1823–5), Le correspondant (1829–30), Berliner allgemeine musikalische Zeitung (1829), Revue européenne (1832), Europe littéraire (1833), Le rénovateur (1833–5), Gazette (later Revue et gazette) musicale (1834–61), Journal des débats (1834–63), Journal des artistes (1834), Monde dramatique (1835), Italie pittoresque (Paris, 1836), Chronique de Paris (1837–8), L'éclair (Brussels, 1842), L'émancipation (1843), L'artiste (1844), Monde illustré (1858–9); see Prod'homme (1904, 1956) and Kapp (1917) — 92, 104, 110, 113, 118, 123, 164–6

BIBLIOGRAPHY

BIBLIOGRAPHIES AND LISTS OF WORKS

J.-G. Prod'homme: 'Bibliographie Berliozienne', *SIMG*, v (1903–4), 622–59

T. Müller-Reuter: *Lexikon der deutschen Konzertliteratur* (Leipzig, 1909)

J. Kapp: *Berlioz: eine Biographie* (Berlin, 1917, 7/1922) [incl. list of writings]

C. Hopkinson: *A Bibliography of the Musical and Literary Works of Hector Berlioz 1803–1869* (Edinburgh, 1951)

J.-G. Prod'homme: 'Bibliographie Berliozienne', *ReM* (1956), no.233, pp.97–147

M. G. H. Wright: *A Bibliography of Critical Writings on Hector Berlioz* (thesis submitted for fellowship of the Library Association, 1967)

B. Jacobson: 'A Berlioz Discography', *High Fidelity*, xix (1969), 56

J. B. Ahouse: *The Centenary Years: a Bibliography of Articles in the Periodical Literature, 1967–71* (El Paso, 1974)

W. Dömling: *Berlioz* (Reinbek bei Hamburg, 1977)

H. J. Macdonald: *Berlioz* (London, 1982)

LETTERS

D. Bernard, ed.: *Correspondance inédite* (Paris, 1879, 2/1879; Eng. trans., 1882 as *Life and Letters of Berlioz*, i)

V. Wilder, ed.: 'Vingt lettres inédites . . . adressées à M. Adolphe Samuel', *Le ménestrel*, xlv (1879), 217–59

'Lettres inédites de Hector Berlioz', *Nouvelle revue*, ii/4 (1880), 801; also pubd separately as *Lettres intimes* (Paris, 1882; Eng. trans., 1882 as *Life and Letters of Berlioz*, ii)

O. Fouque: *Les révolutionnaires de la musique* (Paris, 1882)

La Mara [pseud. of I. M. Lipsius], ed.: *Briefe hervorragender Zeitgenossen an Franz Liszt* (Leipzig, 1895–1904)

——: *Briefe von Hector Berlioz an die Fürstin Carolyne Sayn-Wittgenstein* (Leipzig, 1903)

J. Tiersot: 'Lettres inédites de Berlioz', *RHCM*, iii (1903), 426

'Une page d'amour romantique: lettres à Mme Estelle F.', *Revue bleue*, 4th ser., xix (1903), 417, 457, 484, 513; also in extract

J. Tiersot, ed.: *Hector Berlioz: Les années romantiques: 1819–1842* (Paris, 1904)

J.-G. Prod'homme: 'Nouvelles lettres d'Hector Berlioz', *RMI*, xii (1905), 339–82

——: 'Les lettres de Berlioz à Auguste Morel', *Guide musical*, lviii (1912), 605–710

181

——: 'Lettres inédites de Hector Berlioz', *RMI*, xx (1913), 277

J. Tiersot, ed.: *Hector Berlioz: le musicien errant: 1842–1852* (Paris, 1919)

H. Girard: *Emile Deschamps dilettante* (Paris, 1921)

J. Tiersot: 'Lettres de Berlioz sur Les troyens', *Revue de Paris*, xxi/4 (1921), 449, 749; xxi/5 (1921), 146

'Lettres inédites de Berlioz à Edouard Monnais', *ReM*, iv/6 (1923), 206

'Lettres de Berlioz à sa famille', *Bulletin de l'Académie delphinale* (29 Feb 1924)

S. Dupuis: 'Quelques lettres inédites de Berlioz', *Bulletin de la classe des beaux-arts, Académie des sciences de Belgique*, i–ii (1927), 10

M. Fehr: 'Achtzehn Briefe von Hector Berlioz an den Winterthurer Verleger J. Rieter-Biedermann', *Schweizerisches Jb für Musikwissenschaft*, ii (1927), 90

J. Tiersot, ed.: 'Lettres de musiciens écrites en français: du XVe au XXe siècle', *RMI*, xxxvi (1929), 1, 408; xxxvii (1930), 1; also pubd separately, ii (Turin, 1936)

G. Clarence: 'Lettres inédites à Berlioz', *ReM* (1930), no.104, p.400

S. Ginsburg: 'Correspondance russe inédite de Berlioz', *ReM* (1930), no.104, p.417

J. Tiersot, ed.: *Hector Berlioz: au milieu du chemin: 1852–1855* (Paris, 1930)

'Lettres inédites de Berlioz', *ReM* (1935), no.154, p.164

M. Pincherle: *Musiciens peints par eux-mêmes* (Paris, 1939)

J. Barzun, ed.: *Nouvelles lettres de Berlioz* (New York, 1954)

R. Sietz: *Aus Ferdinand Hillers Briefwechsel*, i (Cologne, 1958)

P. Valléry-Radot: 'Une lettre inédite de Berlioz', *Le fureteur médical*, xxiii (1964), 35

Pisma zarubyezhnïkh muzïkantov uz russkikh arkhivov (Leningrad, 1967)

R. Weeda: 'Vier brieven van Hector Berlioz', *Mens en melodie*, xlv (1969), 177

'Sixty-one Letters by Berlioz', *Adam*, xxxiv (1969), 48–87

P. Citron, ed.: *Correspondance générale: I 1803–1832* (Paris, 1972)

M. Marx-Weber: 'Hector Berlioz: unbekannte Briefe an Peter Cornelius', *Mf*, xxix (1973), 235

P. Citron, ed.: *Correspondance générale: II 1832–1842* (Paris, 1975)

R. P. Locke: 'New Letters of Berlioz', *19th Century Music*, i (1977–8), 71

P. Citron, ed.: *Correspondance générale: III 1842–1850* (Paris, 1978)

P. Bloom: 'Berlioz and Officialdom: Unpublished Correspondence', *19th Century Music*, iv (1980–81), 134

Izbrannïye Pisma, i (Leningrad, 1981)

P. Citron, Y. Gérard, H. Macdonald, eds.: *Correspondance générale: IV 1851–1855* (Paris, 1983)

Bibliography

J.-M. Fauquet: 'Deux lettres inédites d'Hector Berlioz', *Revue musicale de Suisse Romande*, xxxvii (1984), 14

CONTEMPORARY MEMOIRS

W. R. Griepenkerl: *Ritter Berlioz in Braunschweig* (Brunswick, 1843)

J. d'Ortigue: *La musique à l'église* (Paris, 1861)

L. Escudier: *Mes souvenirs* (Paris, 1866), 222–57

E. Reyer: *Notes de musique* (Paris, 1875)

A. Etex: *Les souvenirs d'un artiste* (Paris, 1877)

J. B. Weckerlin: *Musiciana* (Paris, 1877), 319

G. A. Osborne: 'Berlioz', *PMA*, v (1878–9), 60

F. Hiller: *Künstlerleben* (Cologne, 1880), 63–143

A. Barbier: *Souvenirs personnels et silhouettes contemporaines* (Paris, 1883), 230

E. Legouvé: *Soixante ans de souvenirs* (Paris, 1886; Eng. trans., 1893)

M. Maretzek: *Revelations of an Opera Manager in 19th-century America: Sharps and Flats* (New York, 1890/*R*1968)

C. Gounod: *Mémoires d'un artiste* (Paris, 1896)

C. Hallé: *Life and Letters of Sir Charles Hallé* (London, 1896); ed. M. Kennedy (London, 1972)

F. Hueffer: *Half a Century of Music in England (1837–1887)* (London, 1889), 151–234

V. V. Stasov: 'List, Shuman i Berlioz v Rossii', *Severnïy vestnik* (1889), no.7, pt.i, pp.115–57; no.8, pt.i, pp.73–110; Eng. trans., 1968, in *Selected Essays on Music*

C. Saint-Saëns: *Portraits et souvenirs* (Paris, 1900, 3/1909), 3

E. Reyer: 'Berlioz: souvenirs intimes', *Monde musical*, xv (1903), 335

C. Saint-Saëns: 'Souvenirs', *Monde musical*, xv (1903), 335

J.-G. Prod'homme: 'Hector Berlioz jugé par Adolphe Adam', *ZIMG*, v (1903–4), 475

P. Cornelius: 'Hector Berlioz in Weimar', *Die Musik*, iv (1904–5), 159

A. Schloesser: 'Personal Recollections of Franz Liszt and Hector Berlioz', *Royal Academy of Music Club Magazine* (Feb 1911)

J. W. Davison: *From Mendelssohn to Wagner*, ed. H. Davison (London, 1912)

G. B. Shaw: 'A Reminiscence of Hector Berlioz', *Berlioz Society Bulletin* (1973), no.81, p.4; (1974), no.82, p.9

LIFE AND WORKS

F. Clément: *Les musiciens célèbres depuis le seizième siècle jusqu'à nos jours* (Paris, 1868, 3/1878), 514

A. Jullien: *Hector Berlioz: la vie et le combat: les oeuvres* (Paris, 1882)

J. Bennett: *Hector Berlioz* (London, 1883)

A. Jullien: *Hector Berlioz: sa vie et ses oeuvres* (Paris, 1888)

183

E. Hippeau: *Berlioz et son temps* (Paris, 1890)
W. H. Hadow: *Studies in Modern Music*, 1st ser. (London, 1893, 10/1921), 71–146
L. Pohl: *Hector Berlioz' Leben und Werke* (Leipzig, 1900)
A. Hahn and others: *Hector Berlioz: sein Leben und seine Werke* (Leipzig, 1901)
R. Louis: *Hector Berlioz* (Leipzig, 1904)
J.-G. Prod'homme: *Hector Berlioz (1803–1869): sa vie et ses oeuvres* (Paris, 1904, 3/1927)
R. Rolland: 'Berlioz', *Revue de Paris*, xi (1904), 65, 331; repr. in *Musiciens d'aujourd'hui* (Paris, 1908, 6/1914; Eng. trans., 1919)
J. Tiersot: *Hector Berlioz et la société de son temps* (Paris, 1904)
——: 'Berlioziana', *Le ménestrel*, lxx–lxxii (1904–6) [series of articles]
A. Boschot: *L'histoire d'un romantique* (Paris, 1906–13, rev.6/1946–50) [i, *La jeunesse d'un romantique*, 1906, 6/1946; ii, *Un romantique sous Louis-Philippe*, 1908, 6/1948; iii, *Le crépuscule d'un romantique*, 1913, 6/1950]
A. Coquard: *Berlioz* (Paris, 1909)
P.-L. Robert: *Etude sur Hector Berlioz* (Rouen, 1914)
J. Kapp: *Berlioz: eine Biographie* (Berlin, 1917, 7/1922)
P.-M. Masson: *Berlioz* (Paris, 1923)
L. Constantin: *Berlioz* (Paris, 1934)
W. J. Turner: *Berlioz, the Man and his Work* (London, 1934)
J. H. Elliot: *Berlioz* (London, 1938, rev. 4/1967)
E. Lockspeiser: *Berlioz* (London, 1939)
G. de Pourtalès: *Berlioz et l'Europe romantique* (Paris, 1939, rev. 2/1949)
A. Boschot: *Portraits de musiciens* (Paris, 1946)
J. Daniskas: *Hector Berlioz* (Stockholm, 1947; Eng. trans., 1949)
F. Knuttel: *Hector Berlioz* (The Hague, 1948)
J. Barzun: *Berlioz and the Romantic Century* (Boston, 1950; rev. and abridged as *Berlioz and his Century*, New York, 1956, rev. 3/1969)
A. Boschot: *Hector Berlioz: une vie romantique, édition définitive* (Paris, 1951/R1975, 2/1965)
T. Tiénot: *Hector Berlioz: esquisse biographique* (Paris, 1951)
H. Barraud: *Hector Berlioz* (Paris, 1955, 2/1966)
A. A. Khokhlovkina: *Berlioz* (Moscow, 1960)
C. Ballif: *Berlioz* (Paris, 1968)
Berlioz and the Romantic Imagination (London, 1969) [Arts Council Exhibition Catalogue]
S. Demarquez: *Hector Berlioz: l'homme et son oeuvre* (Paris, 1969)

BIOGRAPHICAL STUDIES
E. de Mirecourt: *Berlioz* (Paris, 1856, 3/1856)
O. Fouque: *Les révolutionnaires de la musique* (Paris, 1882)

Bibliography

E. Hippeau: *Berlioz intime* (Paris, 1883, 2/1889)

J. Tiersot: 'Un pèlerinage au pays de Berlioz', *Le ménestrel*, li (1884–5), 345, 353, 361

M. Brenet: *Deux pages de la vie de Berlioz* (Paris, 1889)

E. Istel: 'Berlioz und Cornelius', *Die Musik*, ii (1902–3), 366

E. Closson: 'Hector Berlioz à Bruxelles', *Guide musical*, xlix (1903), 33

H. de Curzon: 'Les débuts de Berlioz dans la critique', *Guide musical*, xlix (1903), 28

F. Grenier: 'Cherubini et Berlioz', *Guide musical*, xlix (1903), 4

M. Kufferath: 'Wagner et Berlioz', *Guide musical*, xlix (1903), 24

R. Sand: 'Berlioz et ses contemporains', *Guide musical*, xlix (1903), 23

C. Maclean: 'Berlioz and England', *SIMG*, v (1903–4), 314

M. Brenet: 'L'amitié de Berlioz et de Liszt', *Guide musical*, l (1904), 595–687

C. Malherbe: 'Une autobiographie de Berlioz', *RMI*, xiii (1906), 506

Le livre d'or du centenaire d'Hector Berlioz (Paris, 1907)

R. Blondel: 'La jeunesse médicale de Berlioz', *Chronique médicale* (1 April 1908)

J.-G. Prod'homme: 'Berlioz anecdotique', *Musica*, vii (1908), 44

J. Tiersot: 'Berlioziana: Berlioz à l'Institut', *Le ménestrel*, lxxvi (1910), 259, 269, 276, 283

——: 'Berlioziana: Berlioz, bibliothécaire du Conservatoire', *Le ménestrel*, lxxvi (1910), 226, 235, 244, 252

——: 'Berlioziana: Berlioz, directeur de concerts symphoniques', *Le ménestrel*, lxxv–lxxvi (1909–10) [series of articles]

E. Dupuy: 'Alfred de Vigny et Hector Berlioz', *Revue des deux mondes*, 6th ser., ii (1911), 837

O. Feis: 'Hector Berlioz, eine pathographische Studie', *Grenzfragen des Nerven- und Seelen-lebens*, xii (1911), 81

J.-G. Prod'homme: 'Une aventure d'amour de Berlioz', *ZIMG*, xiii (1911–12), 137

——: 'Hector Berlioz, bibliothécaire du Conservatoire', *Guide musical*, lix (1913), 783, 803

J. Tiersot: 'Hector Berlioz and Richard Wagner', *MQ*, iii (1917), 453–92

H. Labaste: 'Alfred de Vigny, collaborateur d'Hector Berlioz', *Revue universitaire*, xxix (1920), 369

A. Boschot: *Chez les musiciens* (Paris, 1922)

D. Lazarus: 'Un maître de Berlioz: Anton Reicha', *ReM*, iii/8 (1922), 255

J.-G. Prod'homme: 'Wagner, Berlioz and Monsieur Scribe: Two Collaborations that Miscarried', *MQ*, xii (1926), 359

E. Rey: *La vie amoureuse de Berlioz* (Paris, 1929)

K. Emingerova: 'Hector Berlioz à Prague', *Revue française de Prague* (1933), 167

A. Boschot: 'Berlioz et l'Institut', *Monde français*, xv (1949), 220
A. W. Ganz: *Berlioz in London* (London, 1950)
H. Kühner: *Hector Berlioz: Charakter und Schöpfertum* (Olten, 1952)
G. Court: 'Hector Berlioz and Alfred de Vigny', *ML*, xxxvii (1956), 118
L. Lack: 'Berlioz à Londres', *ReM* (1956), no.233, p.87
M. Wright: 'Humbert Ferrand', *Berlioz Society Bulletin* (1957), no.22, p.5; (1957), no.23, p.3
D. Cairns: 'The Pinch of Snuff', *Berlioz Society Bulletin* (1962), no.40, p.3; (1963), no.41, p.3
M. Wright: 'Berlioz's Physiognomy', *Berlioz Society Bulletin* (1963), no.43, p.3
G. Svyet: 'Gektor Berlioz i Rossiya', *Russkaya Mïsl*, xx (2 March 1967)
V. Donnet: 'Hector Berlioz et la médecine', *Lettres et médecins*, i (1969), 3
V. Donnet and C. Moureaux: 'Le baccalauréat-ès-sciences d'Hector Berlioz', *Marseille médical*, iii (1969), 1
J. Gavot: 'Berlioz à Nice', *Cahiers de l'alpe*, xlvi (1969), 122
L. Guichard: 'Berlioz et Stendhal', *Cahiers de l'alpe*, xlvi (1969), 329
M. Hofmann: 'Berlioz en Russie', *Musica*, clxxix (1969), 28
J. W. Klein: 'Berlioz's Personality', *ML*, l (1969), 15
F. Lesure: 'Le testament d'Hector Berlioz, *RdM*, lv (1969), 219
E. Schenk: 'Berlioz in Wien', *ÖMz*, xx (1969), 217
A. Vander Linden: 'En marge du centième anniversaire de la mort d'Hector Berlioz (8 mars 1869)', *Bulletin des beaux-arts, Académie royale de Belgique*, li (1969), 36–75
R. Weeda: 'Hector Berlioz' bezoecken aan Belgiee', *Mens en melodie*, xxiv (1969), 374
F. de la Sablière: 'Quel père fut Hector Berlioz?', *Revue de Paris*, lxxvii/2 (1970), 94
C. Comboroure: 'Harriet Smithson, 1828–37', *Berlioz Society Bulletin*, (1975), no.86, p.15
K. W. Werth: 'Dating the Labitte Catalogue of Berlioz's Works', *19th Century Music*, i (1977–8), 137
J. Crabbe: *Hector Berlioz, Rational Romantic* (London, 1980)
P. Bloom: 'Berlioz à l'Institut Revisited', *AcM*, liii (1981), 171
J.-M. Fauquet: 'Hector Berlioz et l'Association des artistes musiciens: Lettres et documents inédits', *RdM*, lxvii (1981), 210
P. Raby: *Fair Ophelia* (Cambridge, 1982)
E. F. Jensen: 'Berlioz and Gérard de Nerval', *Soundings*, xi (1983–4), 46
S. Baud-Bory: 'Berlioz et Nathan Bloc', *Revue musicale de Suisse Romande*, xxxvii (1984), 2

GENERAL CRITICAL STUDIES
G. de Massougnes: *Berlioz: son oeuvre* (Paris, 1870, 2/1919)

Bibliography

F. Niecks: 'Hector Berlioz and his Critics', *MT*, xxi (1880), 272, 326

G. Noufflard: *Berlioz et le mouvement de l'art contemporain* (Florence, 1883, 2/1885)

R. Pohl: *Hektor Berlioz: Studien und Erinnerungen* (Leipzig, 1884/*R*1974)

C. Saint-Saëns: *Harmonie et mélodie* (Paris, 1885, 9/1923), 249

A. Jullien: *Musiciens d'aujourd'hui* (Paris, 1892), 1

G. de Massougnes: 'Berlioz et les artistes d'aujourd'hui', *Monde musical*, xv (1903), 339

E. Schuré: 'Le génie de Berlioz', *Guide musical*, xlix (1903), 1

T. S. Wotton: 'Hector Berlioz', *PMA*, xxx (1903–4), 15

——: 'Stray Notes on Berlioz', *ZIMG*, v (1903–4), 395

J. Tiersot: 'Berlioziana', *Le ménestrel*, lxx–lxxi (1904–5) [series of articles]

E. Newman: *Musical Studies* (London, 1905, 3/1914)

Le livre d'or du centenaire d'Hector Berlioz (Paris, 1907)

A. Bruneau: 'L'influence de Berlioz sur la musique contemporaine', *Musica*, vii (1908), 36

A. Jullien: *Musiciens d'hier et d'aujourd'hui* (Paris, 1910), 105–57

F. Weingartner: *Akkorde: gesammelte Aufsätze* (Leipzig, 1912)

M. Emmanuel: 'Berlioz', *Correspondant*, cclxxviii (1920), 237

J. Kapp: *Das Dreigestirn: Berlioz, Liszt, Wagner* (Berlin, 1920)

A. W. Locke: *Music and the Romantic Movement in France* (London, 1920)

C. Koechlin: 'Le cas Berlioz', *ReM*, iii/4 (1922), 118

Berlioz: being the Report of a Discussion held on December 17, 1928 (London, 1929)

J. Tiersot: *La musique aux temps romantiques* (Paris, 1930)

T. S. Wotton: *Hector Berlioz* (London, 1935/*R*1970)

W. Mellers: 'A Prophetic Romantic', *Scrutiny*, vii (1938), 119

F. Farga: *Der späte Ruhm: Hector Berlioz und seine Zeit* (Zurich, 1939, 2/1952)

E. H. W. Meyerstein: 'An Approach to Berlioz', *MR*, ix (1948), 97

R. Collet: 'Berlioz: Various Angles of Approach to his Work', *The Score* (1954), no.10, p.6

L. Guichard: *La musique et les lettres au temps du romantisme* (Paris, 1955)

H. Raynor: 'Berlioz and his Legend', *MT*, xcvi (1955), 414

W. Mellers: *Man and his Music: the Sonata Principle* (London, 1957), 34

J. A. Westrup: 'Berlioz and Common Sense', *MT*, ci (1960), 755

D. Cairns: 'Berlioz and Criticism: Some Surviving Dodos', *MT*, civ (1963), 548

C. Ballif: 'Berlioz aujourd'hui', *Cahiers de l'alpe*, xlvi (1969), 106

D. Cairns: 'Berlioz: a Centenary Retrospect', *MT*, cx (1969), 249

187

H. J. Macdonald: 'Hector Berlioz 1969: zur 100. Wiederkehr seines Todestages', *Musica*, xxiii (1969), 112

J. Barzun: 'Berlioz a Hundred Years After', *MQ*, lvi (1970), 1

M. Guiomar: *Le masque et le fantasme* (Paris, 1970)

E. T. Cone: 'Inside the Saint's Head: the Music of Berlioz', *Musical Newsletter*, i (1971), 3, 16; ii (1972), 19

A. E. F. Dickinson: *The Music of Berlioz* (London, 1972)

P. Heyworth, ed.: E. Newman: *Berlioz: Romantic and Classic* (London, 1972)

B. Primmer: *The Berlioz Style* (London, 1973)

K. K. Reeve: *The Poetics of the Orchestra in the Writings of Hector Berlioz* (diss., Yale U., 1978)

W. Dömling: *Hector Berlioz: Die symphonisch-dramatischen Werke* (Stuttgart, 1979)

J. Barzun: 'The Meaning of Meaning in Music: Berlioz Once More', *MQ*, lxvi (1980), 1

M. Clavaud: *Hector Berlioz, visages d'un masque* (Lyons, 1980)

J. Rushton: *The Musical Language of Berlioz* (Cambridge, 1983)

C. Rosen: 'Battle over Berlioz', *New York Review of Books*, xxxi/7 (1984), 40

SPECIALIZED CRITICAL STUDIES

J. d'Ortigue: *Le balcon de l'Opéra* (Paris, 1833), 295

R. Schumann: 'Aus dem Leben eines Künstlers: Phantastische Symphonie in 5 Abtheilungen von Hector Berlioz', *NZM*, iii (1835), 1–51

J. Mainzer: *Chronique musicale de Paris* (Paris, 1838)

P. Scudo: *Critique et littérature musicales* (Paris, 1850), 14–74

H. von Bülow: 'Hector Berlioz: Benvenuto Cellini', *NZM*, xlvii (1852), repr. in *Ausgewählte Schriften* (Leipzig, 1911), i, 190

L. Kreutzer: 'Le Faust d'Hector Berlioz', *Revue et gazette musicale*, xxi (1854), 389; xxii (1855), 10, 27, 41, 91, 97

F. Liszt: 'Berlioz und seine Haroldsymphonie', *NZM*, xliii (1855), 25–97

E. Hanslick: *Aus dem Concertsaal* (Vienna, 1870/R1971, 2/1896)

A. Jullien: *Goethe et la musique* (Paris, 1880), 126

G. Noufflard: *La Symphonie fantastique de Hector Berlioz: essai sur l'expression de la musique instrumentale* (Florence, 1880)

A. Ernst: 'Wagner corrigé par Berlioz,' *Le ménestrel*, l (1883–4), 348

——: *L'oeuvre dramatique de H. Berlioz* (Paris, 1884)

A. Montaux: 'Berlioz: son génie, sa technique, son caractère; à propos d'un manuscrit autographe d'Harold en Italie', *Le ménestrel*, lvi (1890), 235–85

A. Jullien: *Musiciens d'aujourd'hui: deuxième série* (Paris, 1894), 94

Bibliography

M. Brenet: 'Berlioz inédit: Les francs juges, La nonne sanglante', *Guide musical*, xlii (1896), 61, 81

J.-G. Prod'homme: *La damnation de Faust* (Paris, 1896)

E. Destranges: *Les Troyens de Berlioz: étude analytique* (Paris, 1897)

J.-G. Prod'homme: *L'enfance du Christ* (Paris, 1898)

F. Weingartner: *Die Symphonie nach Beethoven* (Leipzig, 1897; Eng. trans., 1906)

P. Flat: 'Le romantisme de Berlioz', *Guide musical*, xlix (1903), 1

M. Puttmann: 'Hector Berlioz als Gesangkomponist', *NZM*, lxx (1903), 645

T. S. Wotton: 'Einige Missverständnisse betreffs Berlioz', *Die Musik*, iii (1903–4), 358

Neue Musik-Zeitung (3 Dec 1903) [Berlioz no.]

RHCM, iii (1903), 403–46 [Berlioz no.]

RMG (1903), no.49 [Berlioz no.]

A. Boutarel: 'Berlioz und seine "Architecturale Musik" ', *Die Musik*, iii (1903–4), 323

R. Sternfeld: 'Ist der Lélio von Berlioz aufführbar?', *Die Musik*, iii/1 (1903), 373

F. Weingartner: 'Le manque d'invention chez Berlioz', *Guide musical*, xlix (1903), 21

E. Colonne: 'La damnation de Faust', *Musica*, vii (1908), 37

A. Jullien: 'Les Troyens', *Musica*, vii (1908), 43

P. Magnette: *Les grandes étapes dans l'oeuvre de Hector Berlioz, i: La symphonie fantastique (1829–32)* (Liège, 1908)

G. Pioch: 'Berlioz littérateur', *Musica*, vii (1908), 45

J. Tiersot: 'Berlioz symphoniste', *Musica*, vii (1908), 41

A. Boschot: 'La musique religieuse de Berlioz', *Correspondant* (1909), no.237, p.1172

——: *Le Faust de Berlioz* (Paris, 1910, rev. 2/1946)

J. Tiersot: 'Les huit scènes de Faust', *Le ménestrel*, lxxvi (1910), 228, 232, 243

J.-G. Prod'homme: 'Les deux Benvenuto Cellini de Berlioz', *SIMG*, xiv (1912–13), 449

P. J. Bone: *The Guitar and Mandolin* (London, 1914, 2/1954/*R*1972)

E. Newman: 'Rouget de Lisle, "La Marseillaise", and Berlioz', *MT*, lvi (1915), 461

T. S. Wotton: 'The Scores of Berlioz and Some Modern Editing', *MT*, lvi (1915), 651

J. Tiersot: 'Hector Berlioz and Richard Wagner', *MQ*, iii (1917), 453–92

T. Mantovani: *La dannazione di Faust di Ettore Berlioz* (Milan, 1923)

G. Abraham: 'The Influence of Berlioz on Richard Wagner', *ML*, v (1924), 239

K. Isoz: 'Le manuscrit original du "Rákóczy" de Berlioz', *Revue des études hongroises*, ii (1924), 5

G. Servières: 'Pièces inédites relatives aux Troyens', *ReM*, v/10 (1924), 147

J. Tiersot: *La damnation de Faust de Berlioz* (Paris, 1924)

W. W. Roberts: 'Berlioz the Critic', *ML*, vii (1926), 63, 133

T. S. Wotton: 'A Berlioz Caprice and its Programme', *MT*, lxix (1927), 704

F. Bonavia: 'Berlioz', *MMR*, lix (1929), 231

C. Lambert: 'The Isolation of Berlioz: Academic Criticism', *Daily Telegraph* (27 April 1929)

T. S. Wotton: 'Berlioz as Melodist', *MT*, lxx (1929), 808

——: *Berlioz: Four Works* (London, 1929)

J.-G. Prod'homme: 'Berlioz, Musset and Thomas de Quincey', *MQ*, xxxii (1930), 98

F. Baser: 'Hector Berlioz und die germanische Seele', *Die Musik*, xxvi (1933–4), 259

E. Lockspeiser: 'Berlioz and the French Romantics', *ML*, xv (1934), 26

J. Tiersot: 'Schumann et Berlioz', *ReM* (1935), no.161, p.89 [Schumann issue]

D. Tovey: *Essays in Musical Analysis*, iv: *Illustrative Music* (London, 1936/R1972) [on *Harold en Italie*, *Le roi Lear* and the 'Scène d'amour' in *Roméo et Juliette*]

T. S. Wotton: 'Infernal Language: a Berlioz Hoax', *MT*, lxxviii (1937), 209

H. Bartenstein: *Hector Berlioz' Instrumentationskunst und ihre geschichtlichen Grundlagen* (Strasbourg, 1939, rev. 2/1974)

E. Newman: 'Les Troyens', *Opera Nights* (London, 1943), 283–324

T. S. Wotton: 'An Unknown Score of Berlioz', *MR*, iv (1943), 224

R. Dumesnil: *La musique romantique française* (Paris, 1944)

M. Sahlberg: *Berlioz et les russes* (diss., U. of Paris, 1944)

E. Haraszti: '1846–1946: un centenaire romantique: Berlioz et la marche hongroise', *ReM* (1946), suppl.

C. Cudworth: 'Berlioz and Wiertz: a Comparison and a Contrast', *RBM*, vi (1952), 275

F. Noske: *La mélodie française de Berlioz à Duparc* (Paris, 1954; Eng. trans., rev. 2/1970)

M. le Roux: 'Invention de Berlioz', *Domaine musical* (1954), no.1, p.81

Berlioz Society Newsletter (New York, 1954–9)

R. W. S. Mendl: 'Berlioz and Shakespeare', *The Chesterian*, xxix (1955), 95; xxx (1955), 1

J. Barzun: *Energies of Art* (New York, 1956), 281–301

J. Chailley: 'Berlioz harmoniste', *ReM* (1956), no.233, p.15

Bibliography

G. Court: 'Berlioz and Byron and Harold in Italy', *MR*, xvii (1956), 229

A. Espiau de la Maestre: 'Berlioz, Metternich et le Saint-Simonisme', *ReM* (1956), no.233, p.65

G. Favre: 'Berlioz et la fugue', *ReM* (1956), no.233, p.38

L. Guichard: 'Berlioz et Saboly', *ReM* (1956), no.233, p.55

E. Reuter: 'Berlioz mélodiste', *ReM* (1956), no.233, p.31

A. Hammond: 'Benvenuto Cellini', *Opera*, viii (1957), 205

J. W. Klein: 'Les Troyens', *MMR*, lxxxvii (1957), 83

A. E. F. Dickinson: 'Berlioz and The Trojans', *Durham University Journal*, xx (1958), 24

——: 'Music for the Aeneid', *Greece and Rome*, 2nd ser., vi (1959), 129

——: 'The Revisions for "The Damnation of Faust" ', *MMR*, lxxxix (1959), 180

A. Addison: 'Beatrice and Benedict: the German Edition', *Berlioz Society Bulletin* (1960), no.33, p.1

A. Copland: 'Berlioz Today', *Saturday Review*, xliii (27 Aug 1960), 33

P. Friedheim: 'Radical Harmonic Procedures in Berlioz', *MR*, xxi (1960), 282

J. W. Klein: 'Berlioz as a Musical Dramatist', *The Chesterian*, xxxv (1960), 35

C. Wallis: 'Berlioz and the Lyric Stage', *MT*, ci (1960), 358

A. W. G. Court: *Hector Berlioz: the Rôle of Literature in his Life and Work* (diss., U. of London, 1961)

D. Cairns: 'Berlioz Triumphant', *Spectator*, ccxi (1963), 231

R. Collet: 'Berlioz and Shakespeare', *New Statesman*, lxvi (1963), 920

G. S. Fraenkel: 'Berlioz, the Princess and "Les Troyens" ', *ML*, xliv (1963), 249

F. V. Grunfeld: 'The Colossal Nightingale', *Reporter*, xxix (1963), 52

G. Warrack: 'Hector, Thou Sleep'st', *MT*, civ (1963), 896

E. C. Bass: *Thematic Procedures in the Symphonies of Berlioz* (diss., U. of North Carolina, 1964)

D. Cairns: 'Berlioz in Verona', *Music and Musicians*, xii/8 (1964), 15

——: 'Berlioz, the Cornet and the *Symphonie fantastique*', *Berlioz Society Bulletin* (1964), no.47, p.2

A. E. F. Dickinson: 'Berlioz's Rome Prize Works', *MR*, xxv (1964), 163

A. S. Garlington: 'Lesueur, "Ossian" and Berlioz', *JAMS*, xvii (1964), 206

P. Hartnoll, ed.: *Shakespeare in Music* (London, 1964)

G. T. Sandford: *The Overtures of Hector Berlioz: a Study in Musical Style* (diss., U. of Southern California, 1964)

191

J. Warrack: 'Berlioz and the "Theatre of the Mind" ', *The Listener*, lxxii (1964), 738

Bulletin de liaison, Association nationale Hector Berlioz (Paris, 1964–)

R. Hyatt: 'The Earliest Compositions of Berlioz', *Berlioz Society Bulletin* (1965), no.51, p.1; (1965), no.52, p.2

J. Rushton: 'Berlioz's Roots in 18th-century French Opera', *Berlioz Society Bulletin* (1965), no.50, p.3

H. J. Macdonald: 'Berlioz's Self-borrowings', *PRMA*, xcii (1965–6), 27

D. Cairns: 'Berlioz's Epic Opera', *The Listener*, lxxvi (1966), 364

——: 'Hector Berlioz (1803–69)', *The Symphony*, ed. R. Simpson, i (London, 1966), 201–31

A. E. F. Dickinson: 'Berlioz's "Bleeding Nun" ', *MT*, cvii (1966), 584

K. Lucas: *Hector Berlioz and the Dramatic Symphony* (diss., U. of Melbourne, 1966)

H. J. Macdonald: 'The Original Benvenuto Cellini', *MT*, cvii (1966), 1042

E. C. Bass: 'Thematic Unification of Scenes in Multi-movement Works of Berlioz', *MR*, xxviii (1967), 45

E. Gräbner: *Berlioz and French Operatic Tradition* (diss., U. of York, 1967)

L. Guichard: 'Berlioz et Heine', *Revue de littérature comparée*, xli (1967), 5

P. H. Tanner: *Timpani and Percussion Writing in the Works of Hector Berlioz* (diss., Catholic U. of America, 1967)

R. Hyatt: '*Le cheval arabe*, *Beverley*, and *Estelle*', *Berlioz Society Bulletin* (1968), no.60, p.6

H. J. Macdonald: *A Critical Edition of Berlioz's Les Troyens* (diss., U. of Cambridge, 1968)

R. F. Weihrauch: *The Orchestrational Style of Hector Berlioz* (diss., U. of Cincinnati, 1968)

D. Cairns: 'Berlioz and Virgil', *PRMA*, xcv (1968–9), 97

J.-M. Bailbé: *Le roman et la musique sous la monarchie de juillet* (Paris, 1969)

J. Barzun: 'Berlioz in 1969: a Fantasia for Friends Overseas', *Adam*, xxxiv (1969), 33

A. E. F. Dickinson: 'Berlioz's Songs', *MQ*, lv (1969), 329

L. Guichard: 'Berlioz et Stendhal', *Cahiers de l'alpe*, xlvi (1969), 114

H. J. Macdonald: *Berlioz Orchestral Music* (London, 1969)

——: 'Berlioz's Orchestration: Human or Divine?', *MT*, cx (1969), 255

——: 'Hector Berlioz 1969: a Centenary Assessment' *Adam*, xxxiv (1969), 35

——: 'The Colossal Nightingale', *Music and Musicians*, xvii/11 (1969), 24

Bibliography

———: 'Two Peculiarities of Berlioz's Notation', *ML*, l (1969), 25

H. Pleasants: 'Berlioz as Critic', *Stereo Review*, xxiii (1969), 89

J. F. Stuart: *The Dramatic World of Hector Berlioz* (diss., U. of Rochester, 1969)

N. Temperley: 'Berlioz and the Slur', *ML*, l (1969), 388

P. Vaillant: 'Berlioz à la bibliothèque de Grenoble', *Cahiers de l'alpe*, xlvi (1969), 133

J. Warrack: 'Berlioz's *Mélodies*', *MT*, cx (1969), 252

M. Wright: 'Berlioz and Anglo-American Criticism', *Adam*, xxxiv (1969), 93

R. Bockholdt: *Berlioz-Studien* (diss., U. of Munich, 1970; Tutzing, 1979)

R. Covell: 'Berlioz, Russia and the Twentieth Century', *Studies in Music*, iv (1970), 40

J. Day: 'Goetz von Berlichingen and *Les Francs-juges*', *Berlioz Society Bulletin* (1970), no.66, p.6

C. Hopkinson: 'Berlioz and the Marseillaise', *ML*, li (1970), 435

H. J. Macdonald: 'The Labitte Catalogue: Some Unexplored Evidence', *Berlioz Society Bulletin* (1970), no.69, p.5

D. Charlton: 'A Berlioz Footnote', *ML*, lii (1971), 157

E. T. Cone: *Berlioz: Fantastic Symphony* (New York, 1971)

S. Ironfield: *L'art et l'artiste dans les écrits de Berlioz* (diss., U. of Liverpool, 1971)

H. J. Macdonald: 'The Labitte Catalogue: More Evidence', *Berlioz Society Bulletin* (1971), no.70, p.7

B. Primmer: 'Some General Thoughts on Berlioz's Methods', *Berlioz Society Bulletin* (1971), no.71, p.3

———: 'Berlioz and Harmonic Intensification', *Berlioz Society Bulletin* (1971), no.72, p.3

N. Temperley: 'The Symphonie fantastique and its Program', *MQ*, lvii (1971), 593

C. B. Wilson: *Berlioz's Use of Brass Instruments* (diss., Case Western Reserve U., 1971)

E. Gräbner: 'Some Aspects of Rhythm in Berlioz', *Soundings*, ii (1971–2), 18

J.-M. Bailbé: *Berlioz: artiste et écrivain dans les Mémoires* (Paris, 1972)

P. J. Dallman: *Influences and Use of the Guitar in the Music of Berlioz* (diss., U. of Maryland, 1972)

R. Bockholdt: 'Die idée fixe der Phantastischen Symphonie', *AMw*, xxx (1973), 190

H. R. Cohen: *Berlioz and the Opera (1829–1849)* (diss., New York U., 1973)

L. Goldberg: *A Hundred Years of Berlioz's Les Troyens* (diss., U. of Rochester, 1973)

D. McCaldin: 'Berlioz and the Organ', *Organ Yearbook*, iv (1973), 3

R. S. Silverman: 'Synthesis in the Music of Hector Berlioz', *MR*, xxxiv (1973), 346

D. K. Holoman: *Autograph Musical Documents of Hector Berlioz, c.1818–1840* (diss., Princeton U., 1974; Ann Arbor, 1980)

B. Primmer: 'Berlioz and a Romantic Image', *Berlioz Society Bulletin* (1974), no.83, p.4

J. Rushton: 'Berlioz's "Huit scenes de Faust": New Source Material', *MT*, cxv (1974), 471

P. A. Bloom: 'Orpheus' Lyre Resurrected: a tableau musical by Berlioz', *MQ*, lxi (1975), 189

W. Dömling: 'Die Symphonie fantastique und Berlioz' Auffassung von Programmusik', *Mf*, xxviii (1975), 260

D. K. Holoman: 'The Present State of Berlioz Research', *AcM*, xlvii (1975), 31–67

——: 'Reconstructing a Berlioz Sketch', *JAMS*, xxviii (1975), 125

H. Kühn: 'Antike Massen: zu einigen Motiven in *Les Troyens* von Hector Berlioz', *Opernstudien: Anna Amalie Abert zum 65. Geburtstag* (Tutzing, 1975), 141

J. Rushton: 'The Genesis of Berlioz's "La Damnation de Faust" ', *ML*, lvi (1975), 129

R. F. Weihrauch: 'The Neo-Renaissance Berlioz', *MR*, xxxvi (1975), 245

D. Cairns: 'Spontini's Influence on Berlioz', *From Parnassus: Essays in Honor of Jacques Barzun* (New York, 1976)

J. R. Elliott jr: 'The Shakespeare Berlioz Saw', *ML*, lvii (1976), 292

P. Friedheim: 'Berlioz and Rhythm', *MR*, xxxvii (1976), 5–44

D. K.Holoman: 'Berlioz au Conservatoire: notes biographiques', *RdM*, lxii (1976), 289

Romantisme (1976), no.12 [Berlioz no., incl. articles by J. M. Bailbé, D. Cairns, P. Citron, B. Didier, J. Goury, A. Laster]

P. A. Bloom: 'Berlioz and the Critic: *La damnation de Fétis*', *Studies in Musicology in Honor of Otto E. Albrecht* (Kassel, 1977)

J. Rushton: 'Berlioz through the Looking-Glass', *Soundings*, vi (1977), 51

RdM, lxiii (1977) [Berlioz no., incl. articles by H. Barraud, P. A. Bloom, H. R. Cohen, E. W. Galkin, R. P. Locke, H. Macdonald]

P. Bloom: 'A Return to Berlioz's *Retour à la Vie*', *MQ*, lxiv (1978), 354

P. Bloom and D. K. Holoman: 'Berlioz's Music for *L'Europe littéraire*', *MR*, xxxix (1978), 100

S. Ironfield: 'Creative Developments of the "Mal de l'isolement" in Berlioz', *ML*, lix (1978), 33

Bibliography

J. Haar: 'Berlioz and the "First Opera" ', *19th Century Music*, iii (1979–80), 32

T. K. La May: 'A New Look at the Weimar Versions of Berlioz's *Benvenuto Cellini*', *MQ*, lxv (1979), 559

F. Piatier: *Benvenuto Cellini de Berlioz ou le mythe de l'artiste* (Paris, 1979)

E. T. Cone: 'Berlioz's Divine Comedy: the Grande Messe des Morts', *19th Century Music*, iv (1980–81), 3

Silex, xvii (1980) [Berlioz no., incl. articles by J.-V. Richard, G. Rannaud, G. Condé, P. Boulez, D. Lémery, S. Baudo, G. Coutance, B. Didier, M. Clavaud, C. Ballif, J.-F. Héron, G. Duval-Wirth, P. Reliquet, A. Bourmeyster, M. Antoine]

P. Bloom: 'Berlioz and the Prix de Rome of 1830', *JAMS*, xxxiv (1981), 279

J. Rushton: 'Berlioz's Swan-Song: Towards a Criticism of *Béatrice et Bénédict*', *PRMA*, cix (1982–3), 105

C. Berger: *Phantastik als Konstruktion* (Kassel, 1983)

B. Wilt: 'Omkring Berlioz' "Kongelig jakt og storm" ', *Konsertnytt*, xvii/9 (1983), 5

D. K. Holoman: 'Orchestral Material from the Library of the Société des concerts', *19th Century Music*, vii (1983–4), 106

——: 'The Berlioz Sketchbook Recovered', *19th Century Music*, vii (1983–4), 282

T. Tchamkerten: 'Un autographe inédit de Berlioz: Le chant du neuf Thermidor', *Revue musicale de Suisse Romande*, xxxvii (1984), 22

FELIX MENDELSSOHN

Karl-Heinz Köhler

Eveline Bartlitz

CHAPTER ONE

Life

I Family and childhood

On the paternal side of Felix Mendelssohn-Bartholdy's family, his grandfather was the philosopher of the Enlightenment, Moses Mendelssohn (1729–86, originally Moses Dessau), whose characteristic philosophical and literary views played a significant part in the education and thought of young Felix. Together with his friends Gotthold Ephraim Lessing and Friedrich Nicolai, he laid the foundation stone of German national literature. He fought against religious intolerance and the anti-semitic excesses of 18th-century Prussia. His methodology follows classical models in its use of dialogue in order to approach the truth by the dialectic confrontation of opposing viewpoints. Lessing raised an enduring monument to him as the eponymous *Nathan der Weise.*

The civil rights that had extended to Germany after the French Revolution were an essential precondition for the development of the Mendelssohn family; in particular, the attainment of social equality by citizens of the Jewish faith was a consequence of the Enlightenment and the political movements of the second half of the 18th century. Moses Mendelssohn's son Abraham (1776–1835) worked in Paris as a banker in 1803–4; there he met his lifelong companion Lea Salomon (1777–1842), whom he married in 1804. Her grandfather, as financial adviser to Friedrich II, was a factory and property owner who, as one of the most affluent citizens of Berlin, enjoyed special privileges. His family lived in a cultured environment, in which music played a prominent part; Felix's maternal inheritance was especially conducive to his cultural and artistic development. After their marriage Abraham Mendelssohn moved to Hamburg, where he went into partnership with his brother Joseph in a family banking business. There were four children by his marriage to Lea Salomon: Fanny Cäcilie (1805–47), an excellent pianist and accomplished composer who married the painter Wilhelm Hensel; Jakob Ludwig Felix, who was born on 3 February 1809; Rebekka (1811–58), who married Peter Dirichlet, a professor of mathematics; and Paul (1813–74), a financier.

To force England to her knees, Napoleon had decreed a trade blockade of the Continent in 1810. The bankers and merchants of Hamburg, who were particularly hard hit, sought to circumvent the decree by contraband. The French occupying forces then retaliated with a regime of terror, which many of the victims sought to evade by flight. Abraham Mendelssohn and his family were among those who fled to Berlin in 1811 to escape persecution by Maréchal Davout. In 1813 Abraham Mendelssohn equipped a volunteer armed force at his own expense and gave money for a military hospital in Prague. After the victory against the Napoleonic alliance he became a town councillor in Berlin and enjoyed a rapidly advancing social position.

The flowering of Berlin as an intellectual and artistic metropolis, together with the upper middle-class environment of his parental home, formed the essential foundation upon which Felix Mendelssohn's prodigious talent could be developed still further. At first he was educated by his parents. His father taught him arithmetic and French and his mother instructed him in German, literature and the fine arts; no doubt he also received his first piano lessons from her. A family journey to Paris in 1816 (Abraham Mendelssohn had been entrusted by the Prussian government with the collection of French reparations) remained no more than an episode in Felix's artistic development, although during this stay he took piano lessons from Marie Bigot, who was highly esteemed by Haydn and Beethoven.

In the same year Abraham Mendelssohn followed his brother's advice and had his children baptized; he was himself converted to Christianity six years later. This was of course partly to help give real effect to the promised pledge of emancipation, but still more in the genuine conviction of being able to reconcile the moral content and ethical essence of Christianity with the specific inclinations of the Enlightenment philosophy – a reconciliation that was to prove a vital component of Felix's later religious attitude and convictions. As an outward sign of Christian conversion, the family added the name of Bartholdy (after the former owners of a garden site purchased by his mother's elder brother Jakob Lewin Salomon).

In 1819 Felix's education in general subjects and classical languages was entrusted to Carl Wilhelm Ludwig Heyse, a considerable philologist and the father of the poet and short-story writer Paul Heyse. His instruction in the piano had been taken over in 1815 by Ludwig Berger, an admired pianist and teacher of his time, and the nine-year-old Felix made an extremely successful private début with a *Concert militaire* by

15. *Autograph MS of an early composition by Mendelssohn, a piano piece entitled 'Recitativo' (7 March 1820)*

Family, childhood

F. X. Dušek. In 1819 Carl Friedrich Zelter, friend of Goethe and principal of the Berlin Singakademie, began Mendelssohn's instruction in theory and composition. Among his other teachers were the violinist Carl Wilhelm Henning and the landscape painter Johann Gottlob Samuel Rösel. The artistic talents of young Felix were manifest not only in musical achievements, but also in notable drawings and poems. Among the earliest extant attempts at composition is a piano piece entitled *Recitativo* and dated 7 March 1820 (fig.15). In that early year as a composer there followed in quick succession a profusion of small-scale works in which the boy sought to master the world of musical form. His main influences were the contrapuntal techniques of Bach and the Classical style of Mozart. In addition, the earliest Singspiels, *Die Soldatenliebschaft* and *Die beiden Pädagogen*, show a penchant for the dramatic style of Mozart's works of that genre.

The next phase of the young Mendelssohn's development as a composer, up to the end of his 13th year, is marked by an increased mastery of counterpoint and Classical forms, especially sonata form. From June to August 1821 he wrote the G minor Sonata, which was first published posthumously as op.105. The early works, in particular the first Singspiels, were performed in the parental home for the approval and entertainment of invited guests. In November 1821 Zelter took his pupil to see Goethe in Weimar. The resulting relationship with Goethe, which deepened in succeeding years, was of fundamental significance for the young Mendelssohn in that aspect of his creative work related to the Classical period of German literature. Between 1821 and 1830 Mendelssohn paid several substantial

203

visits to Goethe, whose philosophical emphasis on the dynamic and productive aspects of art proved an enriching influence. Mendelssohn for his part increased Goethe's understanding of the music of the Classical period and attempted (though unsuccessfully) to foster an appreciation of Beethoven. Among other Weimar acquaintances were Mozart's pupil Hummel and the writer and music critic Ludwig Rellstab.

A third phase in Mendelssohn's youthful development began soon after his first meeting with Goethe. He began to compose in the larger forms, above all concertos (for example the Piano Concerto in A minor and the D minor Violin Concerto, both of 1822) and preparatory studies for symphonies. Then came larger-scale sacred works (a setting of Psalm lxvi and the D major *Magnificat*). His exceptional dramatic gifts were manifest in the Singspiels *Die wandernden Komödianten* (1821–2) and *Der Onkel aus Boston oder Die beiden Neffen* (1822–3), the spoken dialogues of which are lost. From July to October 1822 the family travelled to Switzerland by a route from Berlin by way of Potsdam, Brandenburg, Magdeburg, Göttingen, Kassel (where Mendelssohn met Ludwig Spohr), Frankfurt am Main (where he met the pianist Aloys Schmitt), Darmstadt, Stuttgart and Schaffhausen into the Alpine region of the Gotthard, by way of Interlaken to Lake Geneva and Lausanne, into the Valais, to the Simplon, thence to Lake Maggiore and back via Berne and Frankfurt to Weimar (and a further meeting with Goethe). Soon afterwards Swiss folksongs were woven into two of the early symphonies. The fruits of the Swiss journey include the C minor Piano Quartet op.1. Immediately Mendelssohn underwent a fourth and final phase in his youthful development as a composer, making rapid progress towards complete maturity and laying the foundation of a personal style. He composed four more string symphonies, the F minor Violin Sonata op.4, the F minor Piano Quartet op.2 and the C minor Symphony op.11, a double concerto for violin and piano and two for two pianos.

Mendelssohn's childhood years were rounded off by further travels with his father, to Silesia in August 1823 and in summer 1824 to Bad Doberan (near Rostock), where he composed the Overture for wind op.24. It fell to Luigi Cherubini to decide Mendelssohn's future as a musician by pronouncing favourably on the Piano Quartet in B minor op.3 (dedicated to Goethe) during a visit to Paris in March 1825 (where Mendelssohn also met or reacquainted himself with several musicians, including Hummel, Kalkbrenner, Moscheles, Rode, Baillot, Boucher, Rossini, Meyerbeer, Liszt and Kreutzer).

II Berlin, 1825–9

In summer 1825 Abraham Mendelssohn acquired the house and grounds at 3 Leipziger Strasse, Berlin. The cultural atmosphere of upper middle-class prosperity was of great significance for Mendelssohn's young manhood. With its theatrical performances, literary readings and regular Sunday concerts, this house developed into the most important salon in Berlin. Among the foremost guests were the natural scientist Alexander von Humboldt and the philosopher Hegel. Here too the young Mendelssohn formed important friendships with Karl Klingemann, later a diplomat; the orientalist Friedrich Rosen; the music editor and critic Adolf Bernhard Marx; the violinist Eduard Rietz; the actor

16. *Felix Mendelssohn: drawing (14 November 1822) by his brother-in-law, Wilhelm Hensel*

Eduard Devrient and his wife the actress Theresa; Ferdinand David, later Mendelssohn's Konzertmeister in Leipzig; Julius Schubring, the theologian who compiled the texts of *St Paul* and *Elijah*; and the philologist Gustav Droysen.

Next to the writings of Jean Paul it was above all the Schlegel translations of Shakespeare and the poetry of Goethe that exerted a decisive influence on Mendelssohn's circle. Possibly Hegel's aesthetic influenced the young Mendelssohn, who later attended his university lectures. In the *Vorlesungen über die Ästhetik* (Berlin, 1836), Hegel stipulated 'that even in instrumental music the composer should devote equal attention to two aspects – musical structure, and the expression of an admittedly indeterminate content'. Inspired by the classical poetry of Goethe, Mendelssohn composed the String Octet op.20 (October 1825), in which his thoroughly individual maturity is recognizable for the first time. On the composer's testimony the scherzo (see fig.25) is indebted to the final lines of the *Walpurgisnacht* from the first part of *Faust*. In 1826 appeared Mendelssohn's most popular work, the overture to Shakespeare's *A Midsummer Night's Dream* op.21; its first public performance was in Stettin in February 1827 at a concert given by Carl Loewe. The last of Mendelssohn's early operas, *Die Hochzeit des Camacho*, was given its première two months later in the Berlin Schauspielhaus (29 April 1827). Intrigues were staged against the performance. The choice of theme, a socio-critical satire based on episodes from the adventures of *Don Quixote*, was ill-suited to Berlin audiences, and the work itself had weaknesses, not least the dramatic construction of Klingemann's libretto; the result was no more than a *succès d'estime*.

In summer 1827 there followed more travels, which took Mendelssohn into the Harz, then to Thuringia and Franconia and to the south of Germany. In Heidelberg he met Justus Thibaut who, in his book *Über die Reinheit der Tonkunst* (1825), had campaigned for a revival of earlier music, especially the work of Palestrina and Handel. Thibaut aroused Mendelssohn's interest in early Italian music.

Mendelssohn's translation of Terence's Latin comedy *The Woman of Andros*, published by his teacher Heyse, secured him a place at the University of Berlin in autumn 1827. There he attended Hegel's lectures in aesthetics, Carl Ritter's in geography and Eduard Gans's on the history of the liberation movement and on the French Revolution. A crisis in the 19-year-old Mendelssohn's artistic development was ended by the composition of the overture *Meeresstille und glückliche Fahrt* op.27, based on two poems by Goethe. The early creative years in Berlin were rounded off by a cantata commemorating the tricentenary of the death of Albrecht Dürer (April 1828) and another celebrating a convocation of natural scientists and doctors summoned by Humboldt.

In the Berlin Singakademie as early as 1819 Mendelssohn had been in contact with the works of Bach, at a time when the large-scale works were considered impossible to perform. Mendelssohn and his friends in the domestic circle had come to know the *St Matthew Passion* from a manuscript copy. A plan for its revival matured and was put before Zelter by Mendelssohn and Eduard Devrient; after some hesitation Zelter agreed. The performance under the direction of the 20-year-old Mendelssohn that followed at the Singakademie on Unter den Linden was a great achievement of musical history, ushering in the modern cultivation of Bach's music. (For an illustration of Mendelssohn's markings in the score, see fig.17.) Two more performances followed in quick succession, the proceeds of which were devoted to social purposes, in particular the founding of two schools of needlework for girls 'of lower social position'.

III Europe, 1829–35
At his parents' insistence Mendelssohn set out to travel abroad for several years. On 10 April 1829 he left Berlin to accept an invitation to London from Klingemann, who was a diplomat there and who, with the pianist Ignaz Moscheles and his wife, introduced him into London salons and social life. Thus the way was prepared for him to appear before the London public in four large-scale concerts. Among other works, his Symphony op.11 and his E major Double Concerto for two pianos were performed (Mendelssohn and Moscheles were the soloists in the latter work); Mendelssohn also played Beethoven's 'Emperor' Concerto. After the concert season he travelled to Edinburgh, where the first ideas for the Scottish Symphony came to him (30 July 1829). On the way to the Scottish highlands he visited Sir Walter Scott in Abbotsford. On 8 August came the stormy crossing by steamship to the island of Staffa with its 'Fingal's Cave', to which Mendelssohn owed the inspiration of his overture *Die Hebriden* op.26 (originally called *Die einsame Insel*). The final stage of the journey was a visit to Coed Du, the estate of the

17. *A page of Mendelssohn's performing edition of Bach's 'St Matthew Passion', showing the end of no.4, with Mendelssohn's tempo, dynamic and phrasing marks in pencil*

210

Taylor family, who were business connections of Mendelssohn's father. The journey back to London was completed by 10 September. Soon afterwards Mendelssohn injured his knee in a coach accident and this delayed his planned return home. After a sojourn at the country home of Thomas Attwood, a former pupil of Mozart, on 7 December he returned to Berlin, where he staged the Liederspiel *Die Heimkehr aus der Fremde* for the silver-wedding anniversary of his parents.

At the beginning of 1830 Mendelssohn, then just 21 years old, was offered the chair of music at the University of Berlin; he declined, however, recommending his friend Adolf Bernhard Marx instead. In winter 1829–30, while at work on the Reformation Symphony, he had had an attack of measles, but by May was ready to begin an Italian journey, which had been suggested by Goethe. (A volume of letters recounts the experiences of this journey in brilliantly entertaining style.) In Weimar Mendelssohn saw Goethe for the last time and played his own works, pieces by Bach and Weber and Beethoven's Fifth Symphony. Early in June he arrived in Munich and then proceeded through Salzburg, Linz, Vienna, Pressburg (for the coronation of Ferdinand V as King of Hungary) via Graz to Venice (9 October). At the end of October Mendelssohn went by way of Florence to Rome, where he met the music collector Fortunato Santini, the papal choirmaster and Palestrina scholar Giuseppe Baini, Berlioz, and the painter and sculptor Wilhelm Schadow.

In Rome Mendelssohn completed the overture *Die Hebriden*, began composing the Italian Symphony and sketched a setting of Goethe's ballad *Die erste Walpurgisnacht*. In April 1831 he travelled with

Schadow to Naples and Pompeii, whose artistic treasures and natural surroundings much impressed him; he was also deeply moved by the poverty of the Neapolitans. At the end of June 1831 he continued his travels by way of Florence and Genoa to Milan (July), where he sought out the company of the pianist Dorothea von Ertmann, an admired friend of Beethoven's. He then travelled through Switzerland, arriving back in Germany in September. On 17 October 1831 he gave a concert in Munich that included the Symphony op.11, the overture to *A Midsummer Night's Dream* and the newly completed G minor Piano Concerto. In Düsseldorf (November) Schadow arranged Mendelssohn's first introduction to the dramatist Karl Immermann, though negotiations for an opera libretto were unsuccessful. Mendelssohn passed the second winter of his great journey in Paris, where he renewed his acquaintance with Baillot, Cherubini and Heine (who had taken part in the revival of the *St Matthew Passion*), met Chopin and the writer Ludwig Börne, and gave successful concerts at the Conservatoire.

Deeply distressed by the news of the death of Goethe (22 March 1832), Mendelssohn made another visit to London in spring 1832; there he learnt of the death of Zelter (15 May). During the concert season Mendelssohn took part in five concerts that included performances of the *Hebriden* and *Midsummer Night's Dream* overtures and the G minor Piano Concerto. At the end of June he returned to Berlin to apply, at his father's urging, to succeed Zelter as director of the Singakadamie. The election was contested and in a ballot among members of the Singakademie Mendelssohn lost to Zelter's deputy, Carl Friedrich Rungenhagen. From

212

November 1832 to January 1833 he gave four concerts in the Berlin Schauspielhaus; in the last of these, *Die erste Walpurgisnacht* was given its first performance.

Early in 1833 Mendelssohn accepted two important invitations. The Philharmonic Society of London had asked him to conduct the Italian Symphony on 13 May, and at the end of that month he was invited to conduct at the Lower Rhine Music Festival in Düsseldorf. In preparation for these events he travelled on 14 April by way of Düsseldorf to London, where the triumphant successes of the previous visits were repeated. A performance of *Israel in Egypt* in Düsseldorf on 26 May 1833 inaugurated a series of Handel oratorio performances in Mendelssohn's own arrangements; they proved an important contribution to Handel's popularity in Germany. On this occasion, stage spectacle was afforded by the background presentation of *tableaux vivants*. After the successful completion of the music festival, which included performances of Beethoven's Pastoral Symphony and *Leonore* Overture no.3, Mendelssohn received a promising commission that required his services in Düsseldorf for two years as city music director, at an annual salary of 600 thaler with three months' leave of absence. His duties consisted in directing Catholic church music and organizing the city music society, which was to give from four to eight concerts a year. A recuperative holiday in England with his father (who had an accident that delayed their return until the end of August) and a short visit to Berlin bridged the gap until he took up his appointment in Düsseldorf. At the beginning of his stay there a meeting with the Crown Prince of Prussia (later

213

King Friedrich Wilhelm IV) resulted in Mendelssohn's nomination as a member of the musical section of the Berlin Academy of the Arts.

At Düsseldorf Mendelssohn concentrated on the cultivation of Handel's oratorios. Within two years five more oratorios were performed in his own arrangements: *Alexander's Feast* (22 November 1833), *Messiah* (Elberfeld, 9 March 1834), the Dettingen *Te Deum* (17 August 1834), *Judas Maccabaeus* (18 December 1834) and *Solomon* (Cologne, May 1835). He also worked in close collaboration with Immermann to found a theatre in Düsseldorf along the lines of Goethe's reforms in Weimar. Increased prices of admission levied in order to prepare exemplary performances provoked a demonstration during the performance of Mozart's *Don Giovanni* under Mendelssohn's direction. After productions of Goethe's *Egmont*, Cherubini's *Les deux journées* and Marschner's *Hans Heiling*, the formal opening of the theatre was on 28 October 1834. Mendelssohn conducted noteworthy works from the opera repertory, including *Oberon*, *Der Freischütz*, *Fra Diavolo*, *Die Entführung aus dem Serail* and *Die Zauberflöte*. But he was plagued by administrative difficulties, which led to an untimely breach with Immermann and ultimately his departure from Düsseldorf. Among the important works he composed during this period are the overture *Die schöne Melusine* op.32, inspired by Conradin Kreutzer's opera (which Mendelssohn saw in Berlin in 1833) and first performed in London on 7 April 1834, and *St Paul*, an oratorio in the tradition of Handel, first performed at the Lower Rhine Music Festival in Düsseldorf on 22 May 1836. The text of the oratorio was freely adapted from the *Acts of the*

Apostles by Mendelssohn's friend Julius Schubring. Otherwise, it was the smaller forms that prevailed – lieder and genre pieces for the piano, appropriately entitled *Lieder ohne Worte* – composed mainly for the salon of the hospitable music-loving family of Woringen, the Düsseldorf city president.

IV Leipzig and the Gewandhaus, 1835–40

At the start of 1835, after successful negotiations with the lawyer Konrad Schleinitz (acting on behalf of Leipzig's town council), Mendelssohn was named the fifth conductor of the Leipzig Gewandhaus orchestra. Following his departure from Düsseldorf after the final concert (2 July 1835) Mendelssohn returned to Berlin, where he witnessed the uprising of 3 August 1835, directed against the despotism of the military regime, and reported its events in detail in a letter to Klingemann. The first concert season began on 4 October with the overture *Meeresstille und glückliche Fahrt* and Beethoven's Symphony no.4. His activities as conductor in Leipzig developed into what must be regarded overwhelmingly the most far-reaching achievement of his life, sustained by his determination to secure a continuing improvement in the standards of orchestral performance, to advance the social position of its members, to plan programmes consisting of items of special historical interest as well as contemporary music, and to invite leading soloists to perform with the orchestra. During the winter concert season, from October to March, some 20 subscription concerts were organized. In addition there were chamber music performances, quartet evenings, and performances of cantatas and oratorios. Mendelssohn acted not only as conductor but also as a soloist, especially as pianist and organist. The contract agreed with the Leipzig town council assured him an annual income of 1000 thaler with permission to leave Leipzig for six months a year. He also received an honorary doctorate from the University of Leipzig.

On 19 November Mendelssohn was shattered by the death of his father. After directing the Lower Rhine Music Festival in Düsseldorf (May 1836) in the first performance of *St Paul*, a performance of Beethoven's Ninth Symphony and the first concert performance of the *Leonore* Overture no.1, Mendelssohn deputized for his friend J. N. Schelble in summer 1836 as director of the Cäcilienverein, an amateur choral society in Frankfurt am Main. During this time he lived with relatives of his Leipzig friend Schleinitz. His hostess, the widow of a minister of the Reformed Church, lived with her two daughters in Frankfurt. After a brief visit to Scheveningen with Schadow in August 1836 he became engaged to one of the daughters, Cécile Charlotte Sophia Jeanrenaud (1817–53), and the wedding was on 28 March 1837 in Frankfurt. On the honeymoon journey (to Freiburg and the Black Forest) and during the following summer sojourn in Bingen am Rhein, he composed the String Quartet in E minor op.44 no.2, the D minor Piano Concerto and a setting of Psalm xlii. After directing the Birmingham Music Festival (19–22 September 1837) in another performance of *St Paul*, Mendelssohn moved to Leipzig with his young wife and took up residence in Lurgensteins Garten. There were three sons and two daughters of this happy marriage.

In the decade of his activity in Leipzig Mendelssohn campaigned for the recognition of forgotten 18th-century works. He introduced Bach's orchestral suites into the civic concert hall. His performance, with Clara Wieck and Louis Rakemann, of Bach's Concerto in D minor for three pianos and orchestra (9 November 1835) was a sensational success. He fostered an increasing interest in Mozart's symphonies, as well as in the symphonies and concertos of Beethoven; Beethoven's Ninth Symphony was performed at the Gewandhaus six times in just ten years. Mendelssohn was also an enthusiastic supporter of Weber's concertos, overtures and operas (scenes from which were frequently given in concert performances). One of the great events of the day was the première of Schubert's C major Symphony (21 March 1839). Schumann had located the manuscript on New Year's Day 1839, at the home of Schubert's brother Ferdinand, and had sent it to Mendelssohn for performance at the Gewandhaus; Mendelssohn conducted it 12 times in all.

To arouse and cultivate the general musical appreciation of his audiences, Mendelssohn began in the winter season 1837–8 the so-called 'historical concerts', a kind of history of music in sound. In four concerts he introduced listeners to music from the time of Bach to their own day: works by Handel, Viotti, Haydn, Cimarosa, Naumann, Righini, Mozart, Salieri, Méhul, B. H. Romberg, Beethoven and the Abbé Vogler. Not only was forgotten music revived but contemporary composers were encouraged. Above all, Schumann owed a large measure of his development and fame to the help and sponsorship of Mendelssohn, who conducted the premières of his first two symphonies (31 March 1841 and 5 November 1846) and Piano Concerto (1 January 1846). Others whose cause he advanced by performing their works were Ferdinand David, Niels Gade, Ferdinand Hiller and Julius Rietz.

Not only notable composers but eminent executants appeared in the Gewandhaus concerts. Clara Schumann gave 21 performances with Mendelssohn as conductor. In March 1840 Liszt gave three special concerts in the Gewandhaus. The programmes contain the names of other well-known performers whom Mendelssohn encouraged: Thalberg, Alexander Dreyschock, Raimund Dreyschock (later leader of the Gewandhaus orchestra), Moscheles, Anton Rubinstein, Vieuxtemps and Joachim, who as a 13-year-old prodigy was recommended by Mendelssohn to London concert promoters. Eminent singers also took part in the concerts: from Sweden came Jenny Lind, who made a triumphantly successful appearance in December 1845 (see fig.18); from England Clara Novello, who sang under Mendelssohn's direction in the winter season 1837–8; and from Leipzig itself Liva Frege, to whom Mendelssohn dedicated some of his songs.

Such programmes, with their emphasis on outstanding works of the past and serious contemporary works, together with the recruitment of outstanding performers, pointed the way to the future. There was also an improved standard of orchestral playing, which Mendelssohn brought about by a radical change in leadership at the outset of his Leipzig activities. The direction of symphonies, previously entrusted to the leader of the orchestra, was taken over by the conductor. An added stimulus to the orchestra's achievement was Mendelssohn's passionate and successful intervention on behalf of the players for increases in salary. With these outstanding achievements as conductor, per-

18. *Programme of a concert, directed by Mendelssohn, held at the Neue Leipzig Gewandhaus on 5 December 1845*

former and music organizer, the years of Mendelssohn's Leipzig activities count among the most significant of his life.

The summer holidays allowed by the contract were usually spent travelling, composing and directing music festivals. In April 1838 Mendelssohn went with his family to Berlin. At the beginning of June he once again undertook the direction of the Lower Rhine Music Festival in Cologne (including a performance of Handel's *Joshua*) and then remained in Berlin for the rest of the summer. The family spent the next summer in Frankfurt am Main, though Felix again assumed the directorship of the Lower Rhine Music Festival in Düsseldorf from 19 to 21 May (this time performing Handel's *Messiah* and Beethoven's Mass in C) and, before the start of the Leipzig concert season, conducting *St Paul*, the D minor Piano Concerto and Beethoven's fifth and seventh symphonies at the Brunswick Music Festival (6–8 September). During the summer recess of 1840 *St Paul* was performed in Weimar on 15 April, and the family spent a two-month holiday in Berlin. But the main event was the Leipzig Festival on 25 June celebrating the 400th anniversary of the invention of the printing press. Mendelssohn was commissioned to write the festival music. His second symphony, the symphonic cantata *Lobgesang* op.52, was performed on 25 June at the Thomaskirche in Leipzig, together with Weber's *Jubel-Ouvertüre* and Handel's Dettingen *Te Deum*. At the North German Music Festival in Schwerin (8–10 July) Mendelssohn once again conducted *St Paul*. On 6 August he inaugurated with a concert of organ music the raising of funds for the erection of a Bach memorial in front of the

Thomaskirche; the unveiling of this memorial was not until 23 April 1843. On 18 September Mendelssohn stayed in London for the sixth time, and at the Birmingham Music Festival (22–5 September) he again conducted the *Lobgesang*.

V Hope and disappointment in Berlin, 1841–2

The death of Friedrich Wilhelm III, King of Prussia, on 7 June 1840 had great significance for Mendelssohn's career. Far-reaching reforms and basic changes were expected from his son and successor, Friedrich Wilhelm IV. Plans for reform were extended to include the arts, thus affecting the musical life of Berlin. The Academy of the Arts was to be reconstituted to include a section for music, as well as for painting, sculpture and architecture. A new conservatory and a new concert organization were to be associated with the music section of the Academy, and the king wanted the energetic Mendelssohn to move to Berlin to be in charge of the whole enterprise. Through the agency of the court official Massow, Friedrich Wilhelm established contact with Paul Mendelssohn, who in November 1840 informed his brother of the ambitious plans. Mendelssohn's protracted correspondence with the king was of little practical result. To begin with, he found himself required to go to Berlin for a year, at three times his salary in Leipzig, but without defined duties except to be in readiness for special employment. He retained his Leipzig position and some of his conducting duties (Ferdinand David acted as deputy conductor in the winter season 1841–2). During summer 1841 he conducted *St Paul* in Weimar (again on 15 April) and in the following months visited Berlin once and Dresden several times, finally arriving at Berlin again at the end of July to take up his new post.

Mendelssohn's first duty was to assist the king in accomplishing the task of ushering in a renaissance of Greek tragedy with incidental music. For this purpose the writer Ludwig Tieck, who had shown himself worthy to take over from Schlegel the task of translating Shakespeare into German, was summoned from Dresden to Potsdam as theatre expert; August Böckh acted as linguistic adviser. The first choice was Sophocles' *Antigone*: it was decided to set the choruses of the classical drama in a modern musical style, with due attention to rules of poetic metre. The composition of the incidental music was completed within 11 days and on 28 October 1841 the first performance took place under Mendelssohn's direction in the small theatre of the Neuer Palais, Potsdam, with an audience of court officials and university professors. On 13 April 1842 *Antigone* was performed at the Berlin Schauspielhaus and was repeated five times in the next three weeks. The setting made a substantial contribution to the popularization of Greek tragedy; the success was sensational and was considered epoch-making.

But despite this success Mendelssohn had not found the field of activity that had been promised him in Berlin. The reforms had not been carried through. The winter season of 1841–2 in Berlin afforded Mendelssohn only one opportunity of public appearance, that of conducting performances of *St Paul* (10 January and 17 February), with only moderate contributions from his collaborators. The king's inability to carry out the promised reforms, deeply rooted in the backward social circumstances in Prussia, were denounced above all in the literature of political reform. In 1843 appeared Bettina von Arnim's book *Dies Jahr gehört dem König*, which advocated the idea of a republican people's monarchy. In 1841 the physician Johann Jacoby published his pamphlet *Vier Fragen* calling for the establishment of a parliament, a share in government for the people, the dismantling of official power and general freedom of thought. Disappointment with the Berlin circumstances induced in Mendelssohn an ever deeper resignation. This is partly reflected in the Scottish Symphony op.56, with which he was occupied in 1841–2, 12 years after the earliest inspiration, and which was first performed at the Gewandhaus under his own direction on 3 March 1842. During winter 1841–2 Mendelssohn travelled to Leipzig many times to fulfil his concert obligations. Summer 1842 was occupied with a further visit to Düsseldorf, where Mendelssohn and Julius Rietz together directed the Lower Rhine Music Festival; Mendelssohn conducted the *Lobgesang* and Handel's *Israel in Egypt* and performed Beethoven's 'Emperor' Concerto at very short notice in place of an indisposed soloist.

At the end of May Mendelssohn paid his seventh visit to England. The climax of his stay in London was the performance of the Scottish Symphony at the Philharmonic Society on 13 June. The composer was twice received by Queen Victoria and the Prince Consort Albert, on 20 June and 9 July, and in gratitude Mendelssohn dedicated the Scottish Symphony to the queen.

On 27 September he returned to Leipzig via Frankfurt am Main and on 2 October conducted the first concert of the season in the Gewandhaus. The next

19. Mendelssohn
with Queen
Victoria and the
Prince Consort:
engraving by
H. Hannal after
G. Durand

day he returned to Berlin. As no attainment of the promised reforms was in prospect, Mendelssohn decided to withdraw from his Berlin obligations, and only the moderating influence of his mother prevented a complete breach with the Prussian king. Mendelssohn waived half his salary. Friedrich Wilhelm IV appointed him general music director and entrusted him with the supervision and direction of sacred music. This meant that Mendelssohn, though still in the king's service, had no binding duties and could instead devote himself entirely to the Gewandhaus orchestra.

The death of his mother on 12 December 1842 caused an emotional crisis that Mendelssohn overcame by undertaking a complete revision of the cantata *Die erste Walpurgisnacht*. The first performance of the new version followed on 2 February 1843 at the Gewandhaus; among those present was Berlioz, who expressed great enthusiasm for the work. Berlioz, whose acquaintance Mendelssohn had made 12 years earlier in Rome, was at the time on a concert tour of Germany during which he had failed to make any great impression; Mendelssohn's selfless assistance in the preparation of the Leipzig concerts helped to make Berlioz recognized for the first time in Germany. Mendelssohn gave similar support to the Danish composer Niels Gade, whose Symphony no.1 was performed on 3 March 1842.

VI The Leipzig Conservatory

On 13 February 1839 the lawyer Heinrich Blumner died, leaving a bequest of 20,000 thaler specifically for 'the founding of a new or the support of an existing all-purpose national institute for the arts or science'. The bequest left the disposition of this sum of money at the discretion of King Friedrich August II of Saxony. Mendelssohn intervened and in April 1840 submitted to Falkenberg, the district director, a petition giving reasons for the need for a music academy in Leipzig. In March 1841 the petition was granted, and after an interval for intensive preparation the new institute was ceremonially opened on 2 April 1843. In the courtyard of the Gewandhaus the city council had a two-storey house built and placed at the disposal of the conservatory. Mendelssohn then devoted himself intensively to the work of organizing the teaching arrangements. His far-reaching ideas included the division into several teaching faculties, the requirement that students practise ensemble playing, the institution of regular, main and auxiliary examinations as well as evening recitals by the students. He laid special emphasis on the maintenance of discipline in teaching and the active encouragement of outstanding talents. He was successful in engaging well-known representatives of special subjects as teachers, among them the choirmaster of the Thomaskirche, Moritz Hauptmann, for harmony and counterpoint; Schumann for the piano, composition and score-reading; Ferdinand David for the violin; and Carl Ferdinand Becker for the organ, music history and theory. Later they were joined by Clara Schumann, Ferdinand Hiller, Gade and Moscheles, who succeeded Mendelssohn in 1847 as director of the conservatory. Among the earliest of the outstanding pupils whom Mendelssohn taught there were the composer and aesthetician Emil Naumann and the violinist and historian Joseph von Wasielewski.

In summer 1843 the boundaries of Mendelssohn's

Berlin duties seemed to be more clearly defined. He was entrusted with the direction of the newly formed male-voice cathedral choir and of the symphony concerts given by the opera orchestra in the 1843–4 season. The king also asked that the series of dramatic productions with incidental music should be continued. Among the plans under consideration, Mendelssohn chose to provide music for *A Midsummer Night's Dream*. In some of the 11 numbers of the incidental music he used themes from the overture written 17 years before. The first complete performance of the incidental music was on 14 October 1843 at the Neuer Palais; the first public performance was four days later at the Berlin Schauspielhaus. Its success was overwhelming. After fulfilling some concert obligations in October and November in Leipzig, Mendelssohn moved with his family to Berlin and occupied apartments in the parental home at 3 Leipziger Strasse.

In his work with the cathedral choir, for which he composed four cantatas to psalm texts (nos.2, 22, 43, 98), Mendelssohn showed a preference for the *a cappella* style of the early Italian masters. His activity as choirmaster, however, was impeded by differences with the choir authorities, especially with the senior court chaplain Strauss. Similar difficulties occurred in planning and rehearsing the ten Berlin symphony concerts, the conducting of which Mendelssohn shared with the opera director Wilhelm Taubert. Mendelssohn concluded the concert season with a performance of Beethoven's Ninth Symphony (27 March 1844) and in April returned to Leipzig. Discouraged again by the circumstances in Berlin, he concentrated on the idea of composing an opera. The plan of collaborating with his

20. *Felix Mendelssohn: portrait (1845) by Eduard Magnus*

friend Eduard Devrient in writing a libretto on a subject from the period of the German peasant revolt took definite shape but was ultimately abandoned. After the summer recess of 1844, when there still seemed no possibility of a change in Berlin, Mendelssohn took steps to release himself almost completely from his obligations there. His salary was further abated on the understanding that for the future he was to continue in the royal service solely as a composer. He furnished two more plays with incidental music, composed early in 1845 but not performed until late autumn. The première of *Oedipus at Colonos* op.93 was on 1 November 1845 at the Neuer Palais; a private performance of Racine's *Athalie* op.74 followed on 1 December at Schloss Charlottenburg. Other similar commissions were refused by Mendelssohn, so that a final breach with the Prussian monarch became inevitable.

In the summer recess of 1844 (May to July) Mendelssohn paid his eighth visit to England, conducting six Philharmonic concerts in performances of his own works as well as those of Bach and Beethoven. In mid July he went to Soden, near Frankfurt am Main, where he relaxed with his family except for a brief interruption by the music festival in Zweibrücken (31 July–1 August); Mendelssohn conducted *St Paul* and *Die erste Walpurgisnacht* and had friendly encounters with the poets Lenau, Hoffmann von Fallersleben and Freiligrath. During this holiday in Soden he composed the Violin Concerto op.64, which was first performed in the Leipzig Gewandhaus on 13 March 1845 with Ferdinand David as soloist and Gade conducting. After a short stay in Berlin, from October to November, during which he conducted a symphony concert on 15 November of Beethoven's Fifth Symphony, *Coriolanus* Overture, and Weber's Overture to *Euryanthe*, he spent the following months in Frankfurt am Main and travelling with his brother and sisters on holiday to Freiburg and Soden.

VII Last years

In August 1845, following a request by the Saxon minister Falkenstein, Mendelssohn resumed his earlier activities at the Gewandhaus, and on 13 August he moved into his last home in Leipzig, at 3 Königstrasse. That summer he completed the Quintet in B♭ op.87. When he resumed the direction of the Gewandhaus concerts on 5 October 1845 he was no longer in full command of his powers. He shared his duties with his deputy Gade. The highpoints of the concert season were the second performance of his Violin Concerto (on 23 October), the concert with the Swedish soprano Jenny Lind (4 December), and the first performance of Schumann's Piano Concerto, with Clara Schumann as soloist (1 January 1846). During the first part of the season Mendelssohn also performed as a pianist, but on medical advice he appeared only as a conductor in the Gewandhaus from early 1846 onwards. At the request of Albert Lortzing he set Schiller's poem *Der Sänger* as a contribution to the festival programme of the Leipzig Schiller Society (10 November). Throughout spring and summer 1846 Mendelssohn's activities were taxing and unremitting. He was committed to several important engagements that involved travelling. He had again been commissioned to take over the direction of the Lower

Rhine Music Festival in Aachen (31 May to 2 June); he
conducted Haydn's *The Creation*, Handel's *Alexander's
Feast* and Beethoven's Fifth Symphony. On 11 June he
was in Liège for the sexcentenary of the feast of
Corpus Christi for which he had set a passage from
Thomas Aquinas's *Lauda Sion Salvatorem*. From 14 to
16 June Mendelssohn took part in the German–Flemish
choral festival in Cologne, for which he had set
Schiller's *An die Künstler* as the festival song op.68.
After a stay in Leipzig he went on his journey to
England, arriving by 18 August 1846 to rehearse the
oratorio *Elijah*, which he had composed in a single
creative outburst in spring and summer 1846 for the
first performance at the Birmingham Music Festival
(26 August 1846; see fig.21). Soon afterwards he
returned to Leipzig via London, Ramsgate, Ostend,
Cologne and Frankfurt am Main, where in the autumn
he undertook a thorough revision of the score of *Elijah*
for publication by Simrock in July of the following
year. During winter 1846–7 he again conducted the
Gewandhaus concerts.

Mendelssohn, who had long been in search of a suitable opera subject, finally settled on the Rhine legend of
the Loreley. In collaboration with Eduard Devrient he
tried to remedy the defects of a libretto by Emanuel
Geibel. This last dramatic composition by Mendelssohn
remained fragmentary. (In 1863 Geibel's text was set by
Max Bruch, though without lasting success.) Nor was he
able to complete his attempt to write a third oratorio,
Christus. On 18 March 1847 he conducted one of his
historical concerts for the last time; and on 2 April, with
a performance of his *St Paul* in the Paulinerkirche, he

21. First performance of Mendelssohn's 'Elijah', at Birmingham town hall, on 26 August 1846: engraving from the 'Illustrated London News' (28 Aug 1846)

conducted for the last time in Leipzig and in Germany. Soon afterwards he visited England for the tenth and last time, conducting several brilliantly successful performances of *Elijah* (16, 23, 28 and 30 April in London, 20 April in Manchester and 27 April in Birmingham). His symphony and chamber concerts also received a tumultuous reception. On 8 May 1847 he began his final journey home, arriving at Frankfurt am Main on 12 May. There he learnt the shattering news of the death of his sister Fanny. In June he sought to recover in Baden-Baden, and at the end of June he travelled with his Berlin relatives via Schaffhausen to Thun. In mid-July he went to Interlaken, where his last great work, the String Quartet in F minor op.80, was composed as a 'Requiem for Fanny'.

On 17 September he returned to Leipzig, where his friends found that he appeared to have changed, looking pale and tired. A week's journey to Berlin to visit his sister's grave brought on a condition of strange perturbation; he became seriously ill and could not conduct even the first of the Gewandhaus concerts. At the end of October he had a slight stroke. Then the attacks became more frequent, leading to a crisis on 3 November. On 4 November, at 9.24 p.m., he died. In his funeral procession to the Paulinerkirche, the pallbearers were Schumann, David, Gade, Hauptmann, Rietz and Moscheles. After the mourning ceremony the coffin was taken by special train to Berlin, where on the following day Mendelssohn was buried near the grave of his sister Fanny in the Trinity Cemetery. Mourning and memorial concerts were given in several cities in Germany and England until the end of the year.

CHAPTER TWO

Works

The catalogue of Mendelssohn's published compositions, which was issued by Breitkopf & Härtel (Leipzig, 3/1882), lists 121 works with opus numbers. The first 72 were published in Mendelssohn's lifetime and numbered according to his wishes; the sequence of these works affords a conspectus of the chronological order only of publication, not of composition. Opp.73–121 were published posthumously between 1848 and 1873, and then appeared in the Gesamtausgabe edited by Julius Rietz (1874–7). In addition some 100 early works, 1821–4, including 13 string symphonies called 'sinfonie', were rediscovered in 1960 and have in part been published in the Leipzig collected edition of Mendelssohn's works. Major collections of Mendelssohn's manuscripts are preserved in Berlin, Oxford, New York, Washington and elsewhere. There is as yet no definitive catalogue of Mendelssohn's work as a whole, though by a careful estimate it comprises about 250 individual items.

Mendelssohn's creative work was moulded by a variety of experiences, trends and influences. It is salutary to distinguish between musical reactions, basic aesthetic concepts and extra-musical influences derived from poetry and emotions which, though subjectively expressed, represent tangible experiences of art, nature, religion, people, society and history. Two principles stand out from the evidence of the compositions themselves, namely the emphasis on clarity and the adherence to the Classical tradition. The young Mendelssohn was guided by Classical and pre-Classical techniques and forms, his first model being the contrapuntal style, in particular the fugal technique of Bach, whose music he copied out as a boy. Among other examples one may point in particular to the convincing chromatic fugal introduction of the Sinfonia no.12 of September 1823. The early work also demonstrates his intensive study of Handel's instrumental technique, which is often apparent in the introductions to the string symphonies, with their typically Handelian rhythms and harmonic progressions.

Mendelssohn's indebtedness to the music of Mozart and his time is no doubt of greater consequence, being perceptible in nearly all the string symphonies written between the ages of 12 and 14, the D minor Violin Concerto of 1822 and the early Singspiels. In the closing bars of *Die beiden Pädagogen* Mendelssohn quoted from the 'Jupiter' Symphony, which he had begun transcribing for piano the same year. The instrumental technique of Beethoven was a formative influence in the early works for full symphony orchestra, especially in the two concertos for two pianos and orchestra. The towering achievement of the young prodigy consists above all in having outgrown a reliance on Classical models and developed a personal style in only four years, between the ages of 11 and 15. His individual stylistic traits include characteristic small-scale rhythmic figures, melodies incorporating a variety of motifs, harmonies intensified by cadential 7th chords, and a freer adaptation of Classical forms. The strikingly individual style of the Mendelssohn of 1825–6, the

235

composer of the String Octet and the overture to *A Midsummer Night's Dream*, proved hardly capable of further change or development during his life. Yet he was able to put his stylistic methods to good effect in a broader thematic application and thus achieved a creative freshness later in such works as the Violin Concerto in E minor (1844) and the Piano Trio in C minor (1845).

The extra-musical influences are shown by a frequent programmatic underlining of musical material. The relationship between content and import can be traced back to the demands of the Hegelian aesthetic, which attributes to music an 'indeterminate content' together with an independently formed musical structure; it is perhaps more than a coincidence that Mendelssohn's acquaintance with the philosopher in his family home was concurrent with the ripening of his style and the interrelationship of his music and the literature with which he was familiar: the Octet, for instance, has a documented connection with lines from the *Walpurgisnacht* scene from the first part of *Faust*. The literary influences were indeed the strongest; they determined the composer's choice and treatment of subject. After Schlegel's Shakespeare translations, Goethe's poetry was the dominant factor, followed by other contemporary German poetry. The influences of art and nature crystallize mainly in the Italian Symphony; emotions from the sphere of personal human relationships are encapsulated in lieder and choruses, and the Scottish Symphony suggests the influence of history as well as of nature. The contrasting feelings of acquiescence and strong protest aroused by his disappointments in Berlin were, on Mendelssohn's own testimony, expressed in

Elijah, and they are also perceptible in the Scottish Symphony.

Neither these various influences nor the composer's own comments about his music permit any definitive classification of Mendelssohn into the mode of thought of any particular period. Least of all is it tenable to call him a Romantic; on the contrary, his affiliations with the 18th century, especially the music of Mozart, make him if anything a neo-classicist.

I   Singspiels and operas

Mendelssohn showed a gift for the dramatic at an early stage in his musical career. For his first Singspiel he chose surprisingly realistic subjects related to his immediate surroundings. True, the trend towards the increasing romanticization of opera subjects inhibited Mendelssohn's later development from the boyhood Singspiel composer who had promised such great success for the future, and this may be reflected in his long, fruitless search for a suitable and above all realistic opera subject. He required of such a subject that it should be drawn from a national event, story or popular legend, and that it should be elevated and serene in style or else should embody a real basic character on the model of *Fidelio*.

The librettist of the early Singspiels, Johann Ludwig Casper (1796–1864), was a friend of the Mendelssohn family. In 1819 he travelled and studied in France, where he came into contact with the critical thought in French vaudeville theatre and translated and made imitations of some of these plays for the young Mendelssohn. *Die Soldatenliebschaft* is set in Countess Elvira's castle and grounds during the French occupation of Spain at the beginning of the 19th century. Mendelssohn obtained striking contrasts by juxtaposing arias, ariettas and ensembles with sharply distinctive rhythms and skilful introduction of different vocal ranges. The psychological relationships are vividly personified in the music, and it is not difficult to recognize the influence of Mozart in the art of characterization as well as in general stylistic features. *Die beiden Pädagogen* also takes place at the beginning of the 19th century, when Pestalozzi's ideas were first beginning to filter through. The play is based on a comedy by Eugène Scribe written in 1817 and called *Les deux précepteurs, ou Asinus asinum fricat*. The musical characterization in this Singspiel is already extended further, especially that of Herr von Robert, a member of the landed gentry, and his rebellious son Carl, whose emotions, views and resolve are depicted in two trios; the typical pedantry of the village schoolmaster is strikingly satirized in the aria 'Probatum est' and in the duet 'Ei, da möchte man ersticken'. And Mendelssohn, then scarcely 12 years old, achieved a masterpiece in a quartet whose middle section incorporates the differences of opinion about the ideas of Pestalozzi and Basedow. The dramatic finale 'Reichen Segen gab der Himmel', with a quartet of soloists, chorus, dance and violin solo, also shows Mendelssohn's striking dramatic talent, even though Mozart's influence is again palpable.

The subject of Mendelssohn's early and only completed opera, *Die Hochzeit des Camacho*, is based on a socio-critical satire from *Don Quixote*. The ballet entr' actes for the wedding celebrations in Act 2, in which Cupid's warriors defeat the champions of wealth, allude to the social criticism underlining the plot, which may well have been one of the causes of the demonstrations staged against the first performance on 29 April 1827. The dramatic construction of the opera (the dialogue of which is attributed to Karl Klingemann) shows undeniable weaknesses – the climax of the action, Basilio's feigned death, is not set to music at all – to which the opera's failure can be attributed.

The Liederspiel *Heimkehr aus der Fremde* op.89 (1829) was written in the simplest of musical styles and remains unrelated to Mendelssohn's work as a whole. Among further opera plans, a work based on Shakespeare's *The Tempest* (in collaboration with Karl Immermann) took shape in 1831 and so in 1834 did *Pervonte* (from a Kotzebue comedy adapted by Klingemann); but neither was completed and no music survives. Nor did plans to write an English opera with James Robinson Planché, the librettist of *Oberon*, come to fruition. The romanticization of opera subjects after 1830 and, most probably, Mendelssohn's premature breach with the operatic stage in Düsseldorf did not permit his operatic and dramatic gifts to achieve their full development; his only late attempt at opera, *Die Loreley* (1847), amounted to no more than a fragment.

II Incidental music

The overture and romance which Mendelssohn composed in Leipzig in 1839 for Hugo's drama *Ruy Blas* came as a consequence of the Berlin commissions. Hugo's tragedy treats of the rise and fall of Ruy Blas in the Spain of Charles II, although the main theme in the tragedy is less the political scene than a lackey–hero's love for his queen. At first Mendelssohn refused the commission (which reached him in March 1839) for an

22. *Durham Cathedral: drawing (24 July 1829) by Mendelssohn*

overture for a performance of the play in aid of the Leipzig theatre's pension fund. Instead he set the romance *Wozu der Vöglein Chöre belauschen fern und nah* for a chorus of six to eight sopranos and strings. He pleaded lack of time to write the overture, but the truth was that he found the play 'entirely repulsive, and beneath all dignity'. But eventually he became unable to endure the accusations that he was taking too much time to compose the overture and despite the heavy burden imposed by his concert duties at the Gewandhaus he finished it in three days, with plenty of time for rehearsal. It was intended as an independent concert overture; its dynamics convey a sense of tension by juxtaposing sharp contrasts between the powerful, thrice-repeated wind chords of the introduction and the main theme with which they are contrasted. With its tragic content, it is numbered among the more significant of Mendelssohn's dramatic compositions.

The incidental music to Sophocles' *Antigone*, the result of a personal commission from Friedrich Wilhelm IV of Prussia, was undertaken in collaboration with the philologist August Böckh and the writer Ludwig Tieck. Classical Greek tragedy is confined to a few actors who describe the action in brief dialogues and a chorus that observes and interprets the events; the ancients used music as a symbol of contemplation. Mendelssohn set the choruses with all the modern musical resources at his command, yet in conformity with the ancient Greek manner (i.e. closely observing the metrical rules of the poetry). For this purpose he followed the precedent of Gluck's operas. The philologist Jacob Christian Donner's translation was chosen as a basis for the adaptation. While this attempt did not

241

escape criticism in philological circles, its aim of popularizing classical Greek tragedy in Berlin was completely successful.

Mendelssohn's most popular work is rightly held to be the overture and incidental music to Shakespeare's *A Midsummer Night's Dream*. He had already written the overture at the age of 17, strongly influenced by Shakespeare's plays in the translations of Schlegel, which he had come to know as a boy through readings or performances in his family home. 17 years later, in 1843, he composed incidental music to the play, 13 numbers in all, as one of the commissions from Friedrich Wilhelm IV; in doing so, he relied to a considerable extent on motifs from the overture. The correspondence of the thematic material in the overture with the various dramatic levels of Shakespeare's play is made clear by verbal correspondences within the incidental music. The following dramatic levels may be identified: (*a*) the figures of popular legend Oberon and Titania, together with (*b*) their subordinate spirits, elves and fairies; (*c*) the realistic dramatic level of the classical pairs of lovers Theseus–Hyppolita (the nobility) and Lysander–Hermia and Demetrius–Helena (the gentry), which is shown as a different world in which the dynamics of decisive complications and conflicts is set in motion; and, as an amusing counterpoint to this level, (*d*) the amateur dramatics of the artisans. In the incidental music Mendelssohn used several themes from his overture as leitmotifs for the dramatic levels. Thus he linked the introductory chords of the overture (ex.1) with the entry of Oberon and Titania in the finale of the incidental music, while the ensuing themes in smaller rhythmic units and played by the upper strings (ex.2) are allotted to the song of the spirits, elves and fairies. A third thematic complex of the overture defines the realistic level of the drama, although only the hunting motif (ex.3) is quoted literally in the incidental music (no.8).

Ex.1 *A Midsummer Night's Dream*, overture

Ex.2 ibid

Ex.3 ibid

Variants of other motifs, however, are easily detectable and are invariably associated with the middle dramatic levels. The main theme of the Wedding March, for example, can be understood as a variant of ex.4, while the chromatically falling line of ex.5*a* finds its application at least as an allusion in the thematic material of no.5, where it is associated with Hermia's renunciation

243

(ex.5*b*). The markedly slower variant of ex.4 in the coda of the overture, which has often been compared with the theme of the Mermaid's Song from *Oberon* (which Mendelssohn most likely did not know at the time he composed the overture), is used in the finale of the incidental music to unite the dramatic levels of the fairy world with the human world. The use of the theme given in ex.6, with its characteristic falling 9th (Bottom's 'hee-haw'), is

entirely unequivocal; it appears in *A Dance of Clowns* (no.11), the dance of the artisan mummers.

These themes and their associated motifs in the exposition of the overture provide a basis for the transparent clarity of the twofold form and structure of the incidental music. The individual themes are changed with each new presentation, transposed to other keys, varied harmonically or contrapuntally, and this musical variability is the essential dynamic element of the work. In its beauty and simplicity it demonstrates Mendelssohn's principles of composition more clearly, perhaps, than any other of his compositions: lucidity in the coordination of intelligible themes, the classic unity of form and structure based on a free use of Classical models, and the connection between music and great poetry, a context in which Mendelssohn's dramatic gifts again become apparent in the construction and interweaving of thematic material.

III Oratorios

In the oratorios of Mendelssohn the musical inspiration is derived for the most part from those of Handel, and content derives from his own code of ethics. The relevance of *St Paul* as an allegory of Mendelssohn's own family history and as a profession of his own religious faith may have been the deciding factors in his choice of subject. The idea of writing an oratorio on the story of the apostle began to mature during his stay in Paris early in 1832, but the plan could not be carried out until after he had made his intensive study of Handel's oratorios in Düsseldorf.

Mendelssohn composed *St Paul* between early 1834 and the spring of 1836. The dramatic tension of the subject is matched musically primarily by the inclusion of large-scale choruses. These are allotted partly to the blind rage of the adherents of the old idolatries and partly to the victorious followers of the new faith. In its attack on intolerance and fanaticism the work proclaims a central doctrine of that faith, 'How lovely are the messengers that preach us the gospel of peace!', which Mendelssohn symbolically set as a spacious choral fugue. He matched the symbolic content of individual situations by distinct but easily comprehensible musical symbols. For example, the descriptive passages treated as recitative are assigned to various voices, and are thus wholly depersonalized, a procedure unthinkable in Bach but frequent in Handel. Similarly, the voice of God is presented impersonally, by four women's voices. Mendelssohn drew freely on Bachian models; the overture is based on the chorale *Wachet auf, ruft uns die Stimme*, and the same theme, treated as a powerful choral movement, dominates the central part of the oratorio. Dramatic choruses, contemplative arias, narrative recitative passages alternate with one another. At pauses in the action, Mendelssohn inserted chorales with simple accompaniments, after the model of Bach's Passions and cantatas.

In summer 1844, after the years of dissension in Berlin (and doubtless motivated by them), Mendelssohn was again seized by the idea of writing an oratorio. This time his subject was the legend of the prophet Elijah (*1 Kings* xvii–xix) and the world of the religious dissension in Israel and Judaea over the divinities Jehovah and Baal. He had written to his librettist Julius Schubring on the subject six years earlier, in a letter of 2 November 1838:

In fact I imagined Elijah as a real prophet through and through, of the kind we could really do with today: strong, zealous and, yes, even bad-tempered, angry and brooding – in contrast to the riff-raff, whether of the court or of the people, and indeed in contrast to almost the whole world – and yet borne aloft as if on angels' wings.

The work begins with a brief prologue announcing the prophecy of drought and the advent of Elijah. Then follows the overture, which begins with the sonorities of lower strings and rises to the brilliance of the upper strings and wind, epitomizing the musical symbolism of the work as a whole. A chorus and a duet, a recitative and an aria followed by a second chorus testify to the people's alarm about the onset of drought and the approach of famine. The excitement of these events is contrasted with the restrained musical evocation of the isolated prophet. The dramatic musical climax of the first part of the oratorio is set on Mount Carmel; the firmness of purpose in Elijah's bass aria is set off against the despairing cries of the chorus, 'Hear and answer, Baal'. The leading part in the action, however, is played by the chorus, which embodies the people just as in Handel's oratorios. Here it provides the symbolism for the climax of this part of the work ('The fire descends'). The second part, containing the indictment, escape and ascension of the prophet, unfolds no less dramatically. There are programmatic elements in the presentation of the tempest, the ocean, the earthquake and the fire during Elijah's sojourn on Mount Horeb. The oratorio ends with the sound of choruses in dramatic triumph.

IV Other vocal works

Mendelssohn's cantatas include occasional works – commissioned compositions and festival music – as well

23. Frontispiece (*second title*) *to the first edition of the full
score of Mendelssohn's 'Elijah', published by Simrock in 1847*

248

as those works which were written independently of commissions and psalm settings for liturgical use. Significant among the commissioned works is the cantata *Begrüssung*, a festival piece composed for the assembly of natural scientists convened in Berlin in 1828 by Alexander von Humboldt, to whom the work is dedicated. The text, which celebrates Man's conquest of the forces of nature, is by Ludwig Rellstab.

Die erste Walpurgisnacht op.60, based on one of the preliminary studies for *Faust*, is an outstanding cantata. Its theme is the old folk custom of greeting May and springtime. In his reply to Mendelssohn's announcement of his intention to set the poem, Goethe observed:

> This poem is in a very real sense highly symbolic in its intention. For it must repeatedly happen in the course of world history that something old, established, tested, and reassuring is probed, harassed and oppressed by the emergence of new ideas and if not destroyed is at least entirely hemmed in and immobilized. The Middle Ages, when hatred could and did evoke such reactions, are here effectively epitomized, and a joyful serene enthusiasm flares out once more in all its pristine fire and brightness.

Mendelssohn was wholly faithful to Goethe's interpretation; he set the two basic ideas of the work by contrasting musical means. The spring festival is presented nobly and with stateliness; but the work as a whole is intensified to a pitch of Bacchantic frenzy in the scene depicting the uproar of the simulated witch-dances that drive away the dogmatic and prejudiced adversaries whose inflexibility threatens their old ways of life. The cantata was composed in 1830 and 1831, and revised in 1843. Berlioz, who heard the revised version, said in his memoirs:

> I was at once quite astounded by the quality of the voices, the responsiveness of the singers, and above all by the grandeur of the work. . . .

One must hear Mendelssohn's music to realize what scope the poem offers a skilful composer. He has made admirable use of his opportunities. The score is of impeccable clarity, notwithstanding the complexity of the writing. Voices and instruments are completely integrated, and interwoven in an apparent confusion which is the perfection of art.

In comparison with the splendid achievement of this highly dramatic work, the settings of psalms for liturgical use show a decline in musical quality. In form and structure they recall Bach's cantatas, but without attaining the dramatic effect of their antecedents. In the Leipzig psalm settings, on the other hand, Mendelssohn was able to extract a rich variety of form and expression from the genre. While Psalm cxiv op.51 is laid out as a through-composed work for double chorus, the other cantatas are framed by opening and final choruses and are mainly dominated by solo voices. The cantatas of opp.78 and 91 were written for the Berlin Cathedral choir, and of necessity were scored for restricted forces. This hampered Mendelssohn and eventually led to disputes with Berlin theologians; the use of the harp in op.91 caused a quarrel because it was a profane instrument.

Among Mendelssohn's best-loved church music is his setting of Psalm c. Occasional pieces such as the early *Magnificat* written for the Berlin Singakademie (1822), the *Te Deum* (1826) or the 'Lauda Sion* (1846) are, however, among his less significant works. Likewise, the secular 'Dürer' Cantata (1828) and the festival chorus *An die Künstler* op.68 (1846) are relatively unimportant.

The often-heard reproach that Mendelssohn was a sentimental Romantic and insipid artist usually derives from misinterpretation of the stylistic resources of his choral songs. Barely half his works in this genre were intended for publication, and were only for social purposes; without exception they were occasional pieces written to grace the social gatherings of choral singers after taxing oratorio performances. The singers went for long walks together and, while resting, sang these choral songs 'in the open air', according to the sub-title of the cycle op.41. A few were commissioned for almanacs or special festivals. At all stages of his life Mendelssohn had occasion to write short community songs, whether in Berlin for the Singakademie, for the participants in the Lower Rhine Music Festival or for the choral singers who helped to give the Leipzig concerts. Among the finest examples is the setting of Goethe's *Frühzeitiger Frühling* (op.59 no.2), with its unobtrusive humming of bees in the middle section, and *Die Nachtigall* (op.59 no.4), with its heartfelt folksong style and magical sonorities. The settings of Eichendorff include the warm *Abschied vom Wald* (op.59 no.3) and the ingenious and fiery *Jagdlied* (op.59 no.6). The choral songs were popular throughout the composer's lifetime, and have lost none of their power to delight; they cannot be considered in the context of the Romantic song cycle, nor should they be, in the words of Ernst Wolff, 'transplanted to the concert hall, where they are like wild flowers in a hot-house'.

Even as a child Mendelssohn turned to the form of art song, but he was 18 before he published his first set of songs, op.8. These are a crystallization of his carefree boyhood years among his friends in the talented Berlin circle. The main literary inspiration in those years came from Shakespeare and Goethe. The masterly style of the setting of Hölty's *Hexenlied* is akin to that of *Die erste Walpurgisnacht* and also the mood of the Octet. Men-

251

delssohn's art songs were not originally intended for the large concert hall but for the small circle of intimate music-making, as found in the family, among friends and in middle-class drawing-rooms in the towns where he was professionally active – Berlin, Düsseldorf and Leipzig; they were the result of special suggestions or occasions. If the op.9 set (1830) remains all too evidently the reflection of personal experience, that is, the emotional expression of young love (no.1, *Frage*, later became a theme of the String Quartet op.13), Mendelssohn strove in his next set (op.19*a*) to create something more than a melodic line with a piano accompaniment; he attempted to enhance the meaning of the poem through the music. The piano accompaniments were themselves made into self-contained character-pieces, a development that led to the *Lied ohne Worte* for solo piano. This connection was underlined by Mendelssohn himself in the dual function of op.19 as lieder or as piano pieces.

Mendelssohn's lieder differ widely in form. Together with virtuoso character-pieces there are simple popular melodies. Among the lyrics are pieces from folksong collections like the *Minnelied* op.34 no.1. Many have even become popular songs in their own right (e.g. *Gruss* op.19*a* no.5, to words by Heine). Mendelssohn favoured the strophic song, and in composing musical material appropriate to the content he shaped the melodic line as well as fashioning a characteristic instrumental accompaniment; the demand for singability was also a factor. In these features Mendelssohn showed a kinship with his predecessors in the Berlin lieder school – Reichardt, J. A. P. Schulz, Zelter and Klein – but at the same time his towering superiority.

V Symphonies

Mendelssohn wrote 13 early symphonies for string orchestra (no.8 in D also exists in a version with wind) and five symphonies for full orchestra, three of which were published in his lifetime (the Italian Symphony and the Reformation Symphony were not published until 1851 and 1868, respectively). The early sinfonias demonstrate the composer's development and his links with tradition perhaps better than any other genre. Those composed as exercises for his teacher Zelter were numbered 1–10; two of them and an isolated movement were left unnumbered but are now commonly referred to as 11–13. The first six are the work of a 12-year-old and clearly show the influence of the Viennese Classical style; the first movements of nos.3 and 4 bear the contrapuntal imprint of Bach and Handel.

The influences of both Bach and Handel, and of the Viennese Classics, were even more clearly perceptible in the later string symphonies, which already show traces of a personal style combined with Classical and pre-Classical traits. The emphasis on counterpoint in the first movement of no.7 is continued in the introduction to no.8, whose second theme is constructed within the limitations of a fugue subject. The addition here of wind instruments indicated the completion of Mendelssohn's first real symphony, but he continued to experiment with string symphony writing. No.9 in C represents a splendid achievement in the mastery of the Classical style. A sprightly virtuoso development of a simple and tuneful first theme determines the course of the first movement (Mozartian antecedents are unmistakable here), and the fairy-like atmosphere of the overture to *A*

Midsummer Night's Dream is anticipated in the scherzo and the finale; the trio of the scherzo is an elaboration of a Swiss folksong. In the scherzo of no.11 another Swiss folksong, *Bin alben e wärti Tächter gsi*, is used, as a memento of the family's summer holiday in Switzerland. No.12 offers a masterly confrontation with the principles of fugue; a complex, chromatically descending fugue subject is used as a means of developing a symphonic form from the contrapuntal stimulus of Bach via the polyphonic world of Mozart to a personal mode of expression. In the symphonic movement in C minor (now referred to as no.13) Mendelssohn similarly sought to unite triple fugue with the technique of symphonic development. This last symphonic study was immediately followed by Mendelssohn's first published symphony, op.11 in C minor (composed in 1824, but only published several years later). In form, structure and melodic invention it is akin to the attempts of the previous year, and its autograph, owned by the London Philharmonic Society (to which the work was dedicated), bears the number 13, thereby linking it to the earlier creative period of the string symphonies.

It is perhaps surprising that some five years elapsed before Mendelssohn, who had meanwhile been strongly influenced by Hegel's ideas, continued his symphonic work. This time he turned to a programmatic vein and composed a work for the celebrations planned in remembrance of the Reformation and commemorating the Augsburg Confession. In the event, the celebrations planned for Berlin did not take place, but the stimulus gave Mendelssohn the occasion to formulate the concept of the Reformation in symphonic terms. Noble in character, the symphony begins with a contrapuntal treatment of the psalmodic incipit D–E–G–F♯, which is also a transposition of the main motif of the finale of Mozart's 'Jupiter' Symphony, and leads to a statement of the so-called Dresden Amen. Powerful trumpet motifs usher in the main theme of the Allegro con fuoco (ex.7), which recalls the opening of Haydn's 'London' Symphony; the movement is based on a contrast of exultation and suffering. In the second movement the cheerful strains of band music, which have been said to be an expression of the simple and powerful feelings of joy, belong to an entirely different world, one suggesting

Ex.7 Reformation Symphony, 1st movt

the countryside and its people. After the sustained recitative movement the finale introduces a set of variations, laid out on broad lines, on Luther's confessional chorale *Ein feste Burg ist unser Gott*. Thus the music is linked with the idea behind the work. The closing movement is also thematically related to the opening of the symphony by the recurrence of the Dresden Amen.

The Italian Symphony was completed in Berlin after Mendelssohn's return from Italy in winter 1832. It was successfully first performed in London on 13 May 1833 and has remained Mendelssohn's most popular symphony. According to his own account, a wide range and variety of impressions were concentrated in it, not only from art and nature but also from the realm of personal experience and contact with the vitality of the Italian people. The guiding main theme of the opening

255

24. *Autograph MS of the opening of Mendelssohn's 'Italian Symphony', first performed in London, 13 May 1833*

Allegro is energetic and concise; two more themes add variety to the musical continuity of the first without detracting from its clarity. The Andante breathes a restrained quietude and nobility; according to Moscheles the composer used the theme of a Czech pilgrim song in this movement. The third movement (Con moto moderato) may have been inspired by his study of Goethe's *Lilis Park*, a humorous poem written for Goethe's friend Lili Schönemann, for he wrote to Fanny on 16 November 1830: 'I want to turn *Lilis Park* into a scherzo for a symphony'. A Neapolitan saltarello forms the basis of the Presto finale.

The *Lobgesang* op.52 (1840) was written for the Leipzig celebrations of the 400th anniversary of the invention of printing. In its external form the work bears a general resemblance to Beethoven's Ninth Symphony. The instrumental opening consists of three movements played without a break, and the opening movement is ceremonial in character, with dotted rhythms in the manner of Handel. The folksong-like theme of the Allegretto breathes relaxed enjoyment. It is followed by a variation movement also with songlike melody (Andante religioso) that concludes the instrumental section, constituting about one third of the work. The rest is a great cantata for solo voices, chorus and orchestra and ends triumphantly with a spacious choral fugue.

The first inspiration for the Scottish Symphony op.56 dates back to 1829, the year of Mendelssohn's first visit to England. His letter of 30 July 1829 from Edinburgh suggests a musical transformation of historical events (he had visited Holyrood, the palace of Mary, Queen of Scots). His mood of resignation soon gave way to optimism, and he laid the work aside as new plans for composition took shape; it was not until after the disappointments of 1841 that his mood predisposed him to resume work on the score. This symphony extends far beyond the programmatic poetic description of landscape and allowed Mendelssohn to attain artistic freedom on a number of emotional levels. It is note-worthy that he added to the tempo indications further directions that stress the meaningful and conceptual side of the work and make his intentions clear, as follows:

Tempo indication	*Character indications*
I. Andante con moto —	Introduction
Allegro un poco agitato —	Allegro agitato
Assai animato	
II. Vivace non troppo	Scherzo assai vivace
III. Adagio	Adagio cantabile
IV. Allegro vivacissimo	Allegro guerriero
	Finale maestoso

The introduction to the first movement is restrained and resigned; the main theme in its Allegro section treats the opening idea in variation style. Dynamic intensifications lead to gripping dramatic scenes in which the experience of nature (depiction of the storm and the sea) mingle with the world of emotions. The mood now declines to melancholy, now builds up a resistance to it, until these conflicts finally give way to the elegiac opening theme. In complete contrast, the cheerful dance-like character of the scherzo is derived from Scottish folk idioms. The third movement is again imprinted with resignation and yearning, strongly prophetic of Brahms. The militant mood of the last movement is new and surprising in Mendelssohn; its purposeful energy, interspersed with folktunes, finally gives way to the sombre opening idea of the introduction, treated once again in variation style, but, released from its elegiac mood, ultimately brought to a triumphant close. In its structure, too, the Scottish Symphony stands apart from the earlier symphonic works, with the individual movements joined to one another much as in the piano concertos and the Violin

Concerto op.64. Thus one of Mendelssohn's least resolved instrumental works may at the same time be regarded as the summit of his achievements as a symphonist.

VI Overtures

Like his symphonies, Mendelssohn's overtures emphasize the programmatic element and they may be regarded basically as one-movement symphonic poems. The main works in the genre were preceded by the Overture for wind instruments op.24 (1824), written for the court orchestra of Bad Doberan near Rostock, and the Trumpet Overture op.101 (1826), so called because of its introductory fanfares; both are in C. In the overture to *A Midsummer Night's Dream* op.21, word-painting had progressed to the point where ideas could be identified with specific musical motifs (see also p.242 above).

The overture *Meeresstille und glückliche Fahrt* op.27 is of central importance, representing the young composer's breakthrough from a crisis in his development. The work is based on two poems by Goethe that describe the external events of nature, the first depicting the obstinate stillness of the sea during a lull, the second the rushing power of waves in motion. As well as this description of natural forces the poems contain allegorical allusion to human behaviour in general: they contrast the unremitting and despairing inertia of the mere observer with the life-enhancing dynamism of the active participant. Mendelssohn depicted the content of the poems in two musically contrasted halves.

A musical interpretation of an experience at sea also influences the course of the B minor overture *Die*

Hebriden. Four years later Mendelssohn composed the overture *Die schöne Melusine* op.32. Grillparzer had adapted the Romantic legend of the mermaid Melusine as an opera subject for Beethoven, who eventually decided against setting it. After Beethoven's death Conradin Kreutzer took up the libretto and set it as an opera commissioned by the Königstädtisches Theater in Berlin. Mendelssohn saw a performance there in spring 1833. He disliked Kreutzer's work but expressed the wish to write his own overture based on the subject; in it he combined a penchant for depicting moods aroused by the sea with the expression of lovers' joy and sorrow, all within the broad framework of the legendary subject.

VI Concertos

Mendelssohn wrote concertos throughout his life – three for the piano, two for two pianos, two for the violin and one for violin and piano. If the early concertos of 1822–3 are, like most of the string symphonies, akin to Viennese Classical models in their form, structure and thematic material, the two double piano concertos of 1823 and 1824 evince an amazing maturity.

Mendelssohn continued his series of larger-scale piano compositions with the Piano Concerto in G minor op.25 (1831); it was conceived completely in his head during the Italian journey, and set down on paper in three days in Munich, where it was first performed in 1831. Although inferior in content to the Italian Symphony, the concerto is planned economically and bears some resemblance to the fantasia form: the first two movements run together and the finale is improvisatory in style. Thus the piece goes beyond the conventional concerto form, a trend that Mendelssohn continued in the next two works in the genre. The Piano Concerto in D minor op.40, written during the honeymoon journey of summer 1837, is dominated by Mendelssohn's joyous emotions of that period and is relatively straightforward; Schumann called it a 'fleeting, carefree gift'. To enhance the virtuoso element in the work, Mendelssohn used techniques introduced by Thalberg, such as transferring the melody to an inner part and decorating it with figurations in both hands.

The summit of Mendelssohn's concerto writing is undoubtedly the Violin Concerto in E minor op.64, which he composed while on a recuperative holiday in Soden near Frankfurt am Main in September 1844. A completely carefree mood permeates the work. The passionately sustained first movement, in sonata form, takes its melodic substance from its urgent initial impulse (ex.8) and a more reflective idea. A transitional section

Ex.8 Violin Concerto, 1st movt, main theme
Allegro molto appassionato

joins the first movement to a songlike Andante. In the last movement Mendelssohn demonstrated the virtuoso possibilities of violin technique, at the same time recalling the atmosphere of *A Midsummer Night's Dream*. But the most significant features of the concerto are formal innovations. Not only are the first two movements joined, but the introduction to the finale (ex.9), beginning on the minor subdominant, recalls the opening theme of the first movement in addition to providing an ingenious link with the finale. The first movement, too, has important formal innovations, notably in the omission of the orchestral exposition and in the placing of the cadenza at the end of the development section and the beginning of the recapitulation. One of the most lyrical of Mendelssohn's instrumental compositions, this concerto stands beside those of Beethoven and Brahms as one of the most significant works in the genre.

Ex.9 Violin Concerto, introduction to finale

VIII Chamber and solo instrumental music

Like his solo works for individual instruments, Mendelssohn's chamber music plays a relatively small part in his output; yet its importance for his development should not be underrated. Even as a boy he set himself impressive tasks. Next to the remarkable Sextet in D op.110 (1824) comes the outstanding Octet in E♭ op.20 for strings (1825), written for the 23rd birthday of the violinist Eduard Rietz, one of the closest friends of Mendelssohn's family circle. After five years' development of his compositional technique Mendelssohn created his own musical idiom in the Octet. The first movement begins with a wide-ranging theme that rises from the depths and soars up to radiant heights, followed by a simpler theme that maintains the forward impulse. The independent character of the movement, marked 'con fuoco', is underlined by its rhythm and dynamics. This contrasts with the simple melody of the Andante, whose folksong-like quality is soon quickened in an enlivened variation form and then immersed in a wonderful stream of animated flowing polyphony. The spirit of the 'Walpurgisnacht Scherzo', as the third movement is sometimes known, is shared and enhanced by the finale with its eight-part fugato, in which the theme of the scherzo returns, interacting with the themes of the finale to bring the work to its Bacchantic close.

Like the string quintets op.18 and op.87, Mendelssohn's string quartets have until recently been unjustifiably neglected. One could say that in the early symphonies he was already trying out string quartet ideas, admittedly with more emphasis on orchestral sound than on the independence of the four parts; his development in quartet writing in the Haydnesque sense, however, was comparatively late. The Quartet op.13 (1827) suggests a reflection of the emotions of an early love affair; the piece develops its thematic substance from a love-song with the text: 'Is it true that you are waiting for me in the arbour by the vine-clad wall?'. The Quartet in E♭ op.12 (1829), on the other hand, suggests a musical confrontation with Beethoven. The initial idea suggests the basic contours (and an inversion) of the themes of Beethoven's 'Harp' Quartet, also in E♭, and these permeate the first movement.

The Three Quartets op.44 (1837–8) are Mendelssohn's masterpieces in the genre, and the second, in E minor, has claims to be considered the best. Its elegiac opening theme gives the first movement its carefree yet noble character; the thematic material is distributed among all four parts, in a filigree technique that Mendelssohn also acquired from Beethoven's quartet style. The short rhythmic units of the scherzo recall the realm of elves and fairies in the works of the early period, particularly the overture to *A Midsummer Night's Dream*. After a songlike Andante comes a playfully passionate finale with many formal contrasts. In Mendelssohn's last quartet, in F minor, grief for his sister Fanny is musically expressed as a pessimistic departure from Classical form. Painful despair, wild yearning, dramatic urgency and spectral unrest dominate the thematic ideas, which suggest no optimistic solution to the problems of the composer, who was himself not far from death.

The piano quartets opp.1–3, all written in childhood, were Mendelssohn's first published works and are comparable to Beethoven's first publication, the Three Piano Trios op.1. The third quartet, in B minor, which decided Mendelssohn's choice of career, is outstanding among the early works, and is dedicated to Goethe.

25. *Autograph MS of the opening of the Scherzo from Mendelssohn's Octet in E♭ op.20, composed in 1825*

265

Together with the Octet, the two late piano trios op.49 in D minor and op.66 in C minor represent the peaks of Mendelssohn's achievement in chamber music. Schumann made a striking comment in his discussion of op.49: 'Mendelssohn is the Mozart of the 19th century, the most illuminating of musicians, who sees more clearly than others through the contradictions of our era and is the first to reconcile them'.

Mendelssohn's solo instrumental works are dominated by piano pieces. Apart from the three large-scale sonatas, a fantasia and a variation set, he cultivated the smaller, more intimate form, derived from the song and produced in profusion from 1830 onwards under the title *Lieder ohne Worte*. Six cycles of these partly lyric, partly virtuoso pieces were published in his lifetime; two were published posthumously. He also wrote virtuoso works, fully exploring the artistic possibilities of the instrument in free form and characterization. The outstanding examples of these are the F$\sharp$ minor Fantasia op.28, to which he gave the title 'Sonate écossaise', and the *Variations sérieuses* in D minor op.54, probably his most successful solo piano work. Building on an ingeniously constructed theme that offers a wide range of harmonic interpretations, he used every device of virtuoso thematic transformation.

Though primarily a pianist, Mendelssohn was also an excellent violinist and organist and wrote equally gratefully for these instruments. The Violin Sonata op.4 (1823) is akin to the early piano quartets in its form and structure though it is certainly overshadowed by the F major Violin Sonata of 1838, which is an immensely virtuoso work, especially the last movement. It is interesting that only one of Mendelssohn's early works

266

belongs to this genre, an F major sonata for the violin written in 1820 as an exercise. The Viola Sonata and the two cello sonatas are also typical of their genres. Lyricism prevails in the earlier cello sonata, op.45 in B♭ (1838). The later sonata, op.58 in D (1843), was written for Paul Mendelssohn and manifests spirited passion and brilliant sonorous effects; it was conceived during the tense period between Mendelssohn's artistic disappointments in Berlin and the wide-ranging opportunities offered in Leipzig (the conservatory was inaugurated in 1843) and, like the Scottish Symphony, communicates a concentrated impression of the dramatic tensions and contradictions through which he lived during those years.

Mendelssohn departed from the Classical tradition in his piano writing, but in his organ Preludes and Fugues op.37 (1837) and the noteworthy organ sonatas op.65 (1844–5) he reverted to the contrapuntal style of Bach, as he did also in the numerous early works for organ, mostly fugal exercises.

CHAPTER THREE

Heritage

For nearly a generation immediately after his death, Mendelssohn's formative influence and popularity continued as an active force in the 19th century. But the memorial tributes of friends and relatives were hardly suited to the process of an objective, scientific and lasting assessment. The Gesamtausgabe edited by Julius Rietz was the most important achievement in 19th-century Mendelssohn scholarship, but the short time that it took to produce (it appeared in its entirety between 1874 and 1877) gives rise to doubts about its textual reliability. The anti-semitic trends that were already perceptible in the middle of the 19th century had a powerfully inhibiting effect on the spread of Mendelssohn's music, and they proliferated in the German writings of the generation that followed Wagner. The main musical factor that inhibited Mendelssohn's popularity was an enhanced Romanticism in performance, with considerable broadening of tempos and fluctuations in phrasing, especially at cadences. This led to interpretations of Mendelssohn's music that tended towards the sentimental, and an exaggeration of chromatic melody and harmony. A misunderstanding of his music, attributable to this sentimentalizing, permeated the music criticism in the later part of the 19th century. This was especially widespread at the turn of the century (in the writings of

Riemann, for example) and has still not been wholly overcome. But many eminent musicians were the most enthusiastic admirers and defenders of Mendelssohn, among them Bülow, Brahms and Reger. Ernst Wolff's documentary biography (1906), which is still unsurpassed, was written 'in order to reaffirm decisively Mendelssohn's greatness and originality, in opposition to those superficial and vindictive judgments to which he . . . is still exposed'.

The racial persecutions of the Fascist era led to the destruction of the Mendelssohn memorial in Leipzig in 1936 and the complete suppression of his music everywhere within Hitler's sphere of influence. An intensification of Mendelssohn studies, which had been recommended by Alfred Einstein in his book *Greatness in Music* as early as 1941, has led to a reappraisal. After World War II there were three commemorative years (1947, 1959 and 1972) with numerous exhibitions that led to a further consideration of Mendelssohn's work and personality. Rediscovery and publication of the early works in the new Leipzig edition has measurably increased the music available for performance and has influenced revivals, as well as more Classically orientated interpretations, of his forgotten mature works. Most significantly, these have enabled a truer, more informed assessment to be made of Mendelssohn's development as an accomplished composer.

WORKS

Editions: *F. Mendelssohn-Bartholdy: Werke: kritisch durchgesehene Ausgabe*, ed. J. Rietz (Leipzig, 1874–7) [R]
Leipziger Ausgabe der Werke Felix Mendelssohn Bartholdys, ed. Internationale Felix-Mendelssohn-Gesellschaft (Leipzig, 1960–) [L]

(printed works published in Leipzig unless otherwise stated)

* – autograph or copy

op.	Title	Genre, Text	Completion	Production	Publication or MS	Edition	
							234, 268
—	Quel bonheur pour mon coeur	dramatic scene	March 1820	Berlin ?15 March 1820	*D-Bds*	—	234
—	Ich, J. Mendelssohn ...	Lustspiel, 3, Mendelssohn	?June 1820	—	frag. *Bds	—	
STAGE							235, 237–45
—	Die Soldatenliebschaft	comic opera, 1, J. L. Casper	Nov 1820	Wittenberg, 28 April 1962	*Bds	—	203, 237–8
—	Die beiden Pädagogen	Singspiel, 1, Casper, after Scribe	c March 1821	Berlin, April 1821 (for J. N. Hummel, with str qt acc.); Berlin, 27 May 1962	L	L v/1	203, 235, 238
—	Die wandernden Komödianten	comic opera, 1, Casper, dialogue lost	1822	rehearsed, Berlin, 8 March 1822	*Bds	—	204
—	Der Onkel aus Boston oder Die beiden Neffen	comic opera, 3, Casper, dialogue lost	Nov 1823	Berlin, 7 Feb 1824	*Bds	—	204
10	Die Hochzeit des Camacho	opera, 2, K. Klingemann, after Cervantes: Don Quixote (dialogue lost)	10 Aug 1825	Berlin, 29 April 1827	vocal score Berlin, 1829	R xv/8	207, 238–9
89	Die Heimkehr aus der Fremde	Liederspiel, 1, Klingemann	1829	Berlin, 26 Dec 1829	Berlin, 1851	R xv/9	211, 239
—	Der standhafte Prinz	incidental music, after Calderón	18 March 1833	Düsseldorf, ?1833	*Bds	—	

op.	Title	Text	Completion	Performance	Publication	Edition	*Bds
—	Trala. A frischer Bua bin i	Schnadahüpferl for Immerman's Andreas Hofer	9 Dec 1833	—		—	—
—	Ruy Blas: Romance (ov.: see orchestral works, op.95, and vocal duets, op.77 no.3)	incidental music, V. Hugo	8 March 1839	—	R	R xviii	239, 241
55	Antigone	incidental music, Sophocles	10 Oct 1841	Potsdam, 28 Oct 1841	vocal score, c1843; full score, 1851	R xv/1	222, 241–2
61	A Midsummer Night's Dream (ov.: see orchestral works, op.21)	incidental music, Shakespeare	1843	Potsdam, 14 Oct 1843	vocal score, 1844; full score, 1848	R xv/4	227, 242–5, 261
93	Oedipus at Colonos	incidental music, Sophocles	25 Feb 1845	Potsdam, 1 Nov 1845	full score, 1852	R xv/3	229
74	Athalie	incidental music, Racine	12 Nov 1845	Berlin-Charlottenburg, 1 Dec 1845	c1849	R xv/2	229
98	Loreley, frag. (finale of Act 1 and Vintners' Chorus)	opera, 3, E. Geibel	1847	Birmingham, 8 Sept 1852	?1852	R xv/10	231, 239

ORATORIOS

245–7

op.	Title	Text	Completion	Performance	Publication	Edition	
36	St Paul (Paulus)	J. Schubring, after the Acts of the Apostles	18 April 1836	Düsseldorf, 22 May 1836	Bonn, 1836	R xiii/1	207, 214, 216, 220, 221, 222, 229, 231, 245–6
70	Elijah (Elias)	Schubring, after I Kings xvii–xix	11 Aug 1846; rev. 1847	Birmingham, 26 Aug 1846; Manchester, London, Birmingham, April 1847	Bonn, 1847	R xiii/2	207, 231, 232, 233, 236, 246–7, 248
97	Christus, inc. (?orig. entitled Erde, Himmel und Hölle)	J. F. von Bunsen, after Matthew, Luke, John, Numbers xxiv	1847	Leipzig, 2 Nov 1854	1852	R xiii/3	231

ORCHESTRAL

op.		
—	Sinfonia no.1, C, str, 1821; L i/1	204, 234, 235, 253–62, 263
—	Sinfonia no.2, D, str, 1821; L i/1	
—	Sinfonia no.3, e, str, 24 Dec 1821; L i/1	
—	Sinfonia no.4, c, str, 5 Sept 1821; L i/1	
—	Sinfonia no.5, Bb, str, 15 Sept 1821; L i/1	
—	Sinfonia no.6, Eb, str, aut. 1821; L i/1	
—	Sinfonia no.7, d, str, ?1821–2; L i/1	253
—	Sinfonia no.8, D, str, 27 Nov 1822, arr. orch, 30 Nov 1822; L i/2	253
—	Violin Concerto, d, str, 1822, ed. Y. Menuhin (New York, 1952); L ii/6	204, 235, 260
—	Piano Concerto, a, str, 1822, *Bds	204, 260
—	Sinfonia no.9, C, str, 12 March 1823; L i/3	205, 253–4
—	Concerto, d, vn, pf, str, 6 May 1823; L i/3	205
—	Sinfonia no.10, b, str, 18 May 1823; L i/3	205
—	Sinfonia no.11, F, str, 12 July 1823; L i/3	205, 235, 254
—	Sinfonia no.12, g, str, 17 Sept 1823; L i/3	205, 209, 235, 260
—	Concerto, E, 2 pf, 17 Oct 1823, perf. Berlin, 14 Nov 1824, edn. (Leipzig, 1960); L ii/4	
—	Sinfonia no.13, c, str, 1 movt only, 29 Dec 1823; Li/3	254
11	Symphony no.1, c, 31 March 1824, perf. Berlin, 14 Nov 1824 (Berlin, 1834); perf., London, 25 May 1829, with arr. of op.20/3 as 3rd movt, *GB-Lbm; R i	205, 209, 212, 254
24	Overture for wind instruments, C, 1st version, c July 1824, rev. ?c Nov 1838 (Bonn, 1839); R vii	205, 259
—	Concerto, Ab, 2 pf, 12 Nov 1824, perf. Stettin, 20 Feb 1827, edn. (Leipzig, 1961); L ii/5	205, 235, 260
101	Overture ('Trumpet Ov.'), C, 4 March 1826, perf. Berlin, 18 April 1828, rev. 10 April 1833, perf. London, 10 June 1833 (1867); R ii	259
21	Ein Sommernachtstraum, ov., after Shakespeare: A Midsummer Night's Dream, 6 Aug 1826, perf. Stettin, 20 Feb 1827 (1832); R ii	207, 212, 227, 236, 242, 253, 259, 264
27	Meeresstille und glückliche Fahrt, ov., after Goethe, c June 1828, perf. Berlin, 8 Sept 1828, rev. March 1834 (1835); R ii	208, 215, 259
—	Kindersymphonie, Dec 1827, lost	
—	Kindersymphonie, Dec 1828, lost	
—	Harmonie Musik, ww, Eb, ?1830, *GB-Ob (copy attrib. Mendelssohn)	
26	Die Hebriden ('Fingalshöhle'), 1st version, Overture zur einsamen Insel, 11 Dec 1830, 2nd version, Die Hebriden, 16 Dec 1850, 3rd version, Isles of Fingal, 20 June 1832; perf. London, 14 May 1832 (1833); R ii	209, 211, 212, 259–60
107	Symphony no.5 'Reformation', D, c May 1830, orig. for 25 June 1830 anniversary of Augsburg Confession, perf. Berlin, 15 Nov 1832 (Berlin, 1868); R i	211, 253, 254–5
25	Piano Concerto no.1, g, Oct 1831 (1833); R viii	212, 260
22	Capriccio brilliant, b, pf, 18 May 1832 (1833); R viii; arr. pf, 18 Sept 1831; *F-Pn	
90	Symphony no.4 'Italian', A, 13 March 1833, perf. London, 13 May 1833 (1851); R i	211, 213, 236, 253, 255–7, 260
32	Die schöne Melusine, ov., after Grillparzer, 14 Nov 1833, perf. London, 7 April 1834 (c1836); R ii	214, 260
29	Rondo brilliant, Eb, pf, 29 Jan 1834 (1835); R viii	
103	Trauermarsch, a, wind, c8 May 1836 (1868); R vii	
40	Piano Concerto no.2, d, 5 Aug 1837 (1838); R viii	
43	Serenade and Allegro giojoso, b/D, pf, 11 April 1838 (Bonn, 1839); R viii	216, 220, 261
—	Symphony, Bb, 1838–9, inc.	
95	Ruy Blas, ov., after V. Hugo, 8 March 1839, perf. Leipzig, 31 March 1839 (c1851); R ii	241
52	Symphony no.2 ('Lobgesang'), symphony-cantata, Bb, last movt with solo vv, chorus, org, perf. Leipzig, 25 June 1840, rev. 27 Nov 1840 (1841); R xiv/A2	220, 221, 223, 257
108	March, D, April 1841, perf. Dresden, 1841 (1868); R iii	
56	Symphony no.3 'Scottish', a, 20 Jan 1842, perf. Leipzig, 3 March 1842 (1843); R i	209, 223, 236, 237, 257–9, 267
—	Concerto, e, pf, 1842–4, inc. *GB-Ob	
64	Violin Concerto, e, 16 Sept 1844, perf. Leipzig, 13 March 1845 (1845); R iv	229, 230, 236, 258–9, 261–2
—	Symphony, C, 1844–5, inc., *Ob	

CHAMBER

op.		
—	Allegro, C, vn, pf, c1820, *Ob, ed. R. L. Todd (1983)	
—	Andante, d, vn, pf, c1820, *Ob, ed. R. L. Todd (1983)	
—	Movement, g, vn, pf, c1820, *Ob, ed. R. L. Todd (1983)	
—	Theme and Variations, C, vn, pf, c1820, *Ob, ed. R. L. Todd (1983)	
—	Trio, c, vn, va, pf, 9 May 1820, ed. in J. A. McDonald (1970), 336	262–7

[PIANO SOLO and other works]

Cat.	Work	Pages
109	Lied ohne Worte, D, vc, pf, ? c Oct 1845 (1868); R ix	233, 264
80	String Quartet no.6, f, Sept 1847 (1850); R vi	
81/1	Andante, E, str qt, c Aug 1847 (1850); R vi	
81/2	Scherzo, a, str qt, c Aug 1847 (1850); R vi	
—	Theme, A, str qt, inc., *GB-Ob	
—	Adagio, Allegro molto, D/d, vn, pf, inc., *D-Bds	

PIANO SOLO
(215, 252, 266)

Cat.	Work	Pages
—	Theme and Variations, D, c1820, *GB-Ob, ed. R. L. Todd (1983)	
—	Four little pieces, c1820, *Ob: G, g (canon), G, g (canon), ed. R. L. Todd (1983)	
—	Andante, F, 1820, *D-Bds	
—	Piano piece, e, 1820, *Bds	
—	Two little pieces, 1820, *Bds: Andante, C, Presto, C [?duet]	
—	Two little pieces, 1820, *Bds: Largo, d, untitled, f	
—	Five little pieces, 1820, *Bds: Allegro, C, Allegro, g, Andante, A, untitled, b, untitled, a	
—	Largo–Allegro, c, 1820, *Bds	202, 203
—	Recitativo ('Largo'), d, 7 March 1820, *Bds; version with str, c22 April 1820	
—	Sonata, f, 1820, *Bds	
—	Sonata, a, 12 May 1820, *Bds	
—	Presto, c, 1 July 1820, *Bds	
—	Sonata, e, 13 July 1820, *Bds	
—	Two studies, d, a, 28 Dec 1820, *Bds	
—	Allegro, a, 5 Jan 1821	
105	Sonata, g, 18 Aug 1821 (1868); R xi/3	203
—	Largo–Allegro molto, c/C, ?1821–2, *Bds	
—	Three Fugues, d, d, b, ?1822, *Bds	
—	Allegro, d, 19 Feb 1823, *Bds	
—	Sonata, bb, 27 Nov 1823, ed. R. L. Todd (1981)	
—	Fantasia ('Adagio'), c, ?1823, *Bds	
—	Capriccio, Eb/eb, ? c1823–4, ded. L. Heidemann, *GB-Ob	
—	Prestissimo, f, 19 Aug 1824, *Ob	
5	Capriccio, ff, 23 July 1825 (Berlin, 1825); R xi/1	
—	Capriccio, c#, 5 Jan 1826, *D-Bds	
—	Vivace, c, 29 Jan 1826, *GB-Ob	
—	Andante and Canon, D, ? c Jan 1826, *US-NYpm	

[Chamber works]

—	Minuet, G, vn, pf, 3 Dec 1820, *D-Bds	
—	Presto, F, vn, pf, 1820 *Bds	
—	Violin Sonata, F, 1820, ed. R. Unger (1977)	267
—	Several Fugues, str qt, Jan–May 1821, *Bds	
1	Piano Quartet, d, 1822, ed. in J. A. McDonald (1970), 372	204, 264
—	Piano Quartet no.1, c, 18 Oct 1822 (Berlin, 1823); R ix	205, 266
4	Violin Sonata, f, 3 June 1823 (Berlin, 1825); R ix	205, 264
2	Piano Quartet no.2, f, 3 Dec 1823 (Berlin, 1825); R ix	267
—	Viola Sonata, c, 14 Feb 1824, edn. (Leipzig, 1966)	
110	Clarinet Sonata, Eb, 17 April ?1824, ed. E. Simon (1951)	263
3	Sextet, D, vn, 2 va, vc, db, pf, 10 May 1824 (1868); R ix	205, 264
—	Piano Quartet no.3, b, 18 Jan 1825, perf. Weimar, May 1825 (Berlin, 1825); R ix	207, 236, 251, 263, 265, 266
20	Octet, Eb, 4 vn, 2 va, 2 vc, 15 Oct 1825, parts pubd (1833); R v	263
18	Quintet no.1, A, 2 vn, 2 va, vc, 1st version with Minuetto, f#, 31 March 1826, *US-NYpm, 2nd version with Intermezzo, 23 Feb 1832 (Bonn, 1833); R v	
13	String Quartet no.2, a, 26 Oct 1827 (1830); R vi; L iii/1	252, 263–4
81/4	Fugue, Eb, str qt, 1 Nov 1827 (1850); R vi	
17	Variations concertantes, D, vc, pf, 30 Jan 1829 (Vienna, ?1830); R ix	
12	String Quartet no.1, Eb, 14 Sept 1829 (1830); R vi, L iii/1	264
—	The Evening Bell, harp, pf, Nov 1829, pubd in *Musical Haunts in London* (London, 1830)	
113	Concert Piece, f/F, cl, basset-hn, pf, 30 Dec 1832 (Offenbach, 1869), arr. orch, 6 Jan 1833; R vii [pf version]	
114	Concert Piece, d, cl, basset-hn, pf, 19 Jan 1833 (Offenbach, 1869); R vii	
—	Assai tranquillo, b, vc, pf, 25 July 1835, facs. in Sietz (1962)	
44	String Quartets nos.3–5 (1839): D, 24 July 1838; e, 18 June 1837, perf. Leipzig, 28 Oct 1837; Eb, 6 Feb 1838; R vi, L iii/2	216, 264
—	Violin Sonata, F, 15 June 1838, ed. Y. Menuhin (New York, 1953)	266
45	Cello Sonata no.1, Bb, 13 Oct 1838 (1839); R ix	267
49	Piano Trio no.1, d, early version, 18 July 1839; 2nd version, 23 Sept 1839 (1840); R ix	266
58	Cello Sonata no.2, D, c June 1843, perf. Leipzig, 18 Nov 1843 (1843); R ix	267
81/3	Capriccio, e, str qt, 5 July 1843 (1850); R vi	
66	Piano Trio no.2, c, c April 1845 (1846); R ix	236, 266
87	Quintet, no.2, Bb, 8 July 1845 (1850); R v	230, 263

6	Sonata, E, 22 March 1826 (Berlin, 1826); R xi/1	266
7	Sieben Charakterstücke, (Berlin, 1827): e, 6 June 1826, b, 17 July 1824, D, A, 4 June 1826, A, e, E; R xi/1	
—	Fugue, E♭, 11 Sept 1826, *GB-Ob	
15	Fantasia, E, on 'The Last Rose of Summer', ?1827 (Vienna, ? c1831); R xi/1	
106	Sonata, B♭, 31 May 1827 (1868); R xi/3	
—	Fugue, e, 16 June 1827, added to Prelude of 13 July 1841	
119	Scherzo, b, 12 June 1829, pubd in Berliner allgemeine musikalische Zeitung, vi (1829); R xi/1	
—	Perpetuum mobile, C, ? c June 1829, ? ded. I. Moscheles (1873); R xi/3	
16	Trois fantaisies ou caprices (Vienna, ? c1831): a, 4 Sept 1829, e, 13 Nov 1829, E, 5 Sept 1829; R xi/1	
—	Andante con moto, A, 3 June 1830, ded. O. von Goethe, *D-DUK	
—	Andante, A, 13 June 1830, *D-Bds	
14	Rondo capriccioso, E, 13 June 1830 (Vienna, c1830); early version, étude, 4 Jan 1828, ded. J. Kohlreif, *US-NYpm; R xi/1	
19b	Lieder ohne Worte, i, orig. pubd as Original Melodies for the Pianoforte (London, 1832); E, a, 11 Dec 1830, 'Jägerlied', A, A, 14 Sept 1829, f♯, Venetianisches Gondellied, g, 16 Oct 1830; R xi/4	266
—	Con moto, A, 3 Nov 1831, *US-NYpm	
28	Fantasia ('Sonate écossaise'), f♯, 29 Jan 1833 (Bonn, 1834); R xi/1	266
30	Lieder ohne Worte, ii (Bonn, 1835): E♭, b♭, 30 Sept 1830, E, b, 30 Jan 1834, D, 12 Dec 1833, Venetianisches Gondellied, f♯; R xi/4	266
33	Three Caprices (1836): a, 9 April 1834, E, 12 Sept 1835, b♭, 25 July 1833; R xi/2	
—	Scherzo a capriccio, f♯, 29 Oct 1835, pubd in L'album des pianistes (Paris, 1836); R xi/1	
—	Study, f, 13 March 1836, pubd in Moscheles and Fétis's Méthode des méthodes de piano (Paris, ?1840)/R1973; Eng. trans., 1841); R xi/1	
—	Andante, A♭, 27 June 1836, *D-Bds	
—	Lied, f♯, 16 Oct 1836, *Bds, arr. 2vv, pf as Herbstlied op.63 no.4	
—	Prelude, f, 13 Nov 1836, *Bds	
104a	Three Preludes (1868): B♭, 9 Dec 1836, b, 12 Oct 1836, D, 27 Nov 1836; R xi/3	266
35	Six Preludes and Fugues, 9 Jan 1837 (1837): e/E; D, prelude 8 Dec 1836; b, fugue 21 Dec 1832: A♭, fugue 6 Jan 1835; f, prelude 19 Nov 1836, fugue 3 Dec 1834; B♭, prelude 3 Jan 1837, fugue 27 Nov 1836; R xi/2	
38	Lieder ohne Worte, iii (Bonn, 1837): E♭, c, 29 March 1836, E, 2 Jan 1835, A, a, 6 April 1837, Duetto, A♭, 27 June 1836; R xi/4	
—	Gondellied ('Barcarole'), A, 5 Feb 1837 (Dresden, 1838), pubd as suppl. to NZM, xiv (July 1841); R xi/1	
—	Allegretto, A, 22 April 1837, facs. in J. Petitpierre (1937)	
118	Capriccio, E, 11 July 1837 (1872); R xi/3	
117	Allegro, e, ?1837 ? ded. F. W. Benecke (1872); R xi/3	
—	Andante cantabile and Presto agitato, B, 22 June 1838, pubd in Musikalisches Album (1839); R xi/1	
104b	Three Studies (1868): bb, 9 June 1836, F, 21 April 1834, a, ?1838; R xi/3	
—	Lied ohne Worte, f♯, 5 April 1839	
53	Lieder ohne Worte, iv (Bonn, 1841): A♭, copy 7 March 1839, E♭, 24 Feb 1835, g, 14 March 1839, Abendlied, F, 1 May 1841, Volkslied, a, 30 April 1841, A, 1 May 1841; R xi/4	266
—	Prelude and Fugue, e, pubd in Notre temps (Mainz. 1842): prelude 13 July 1841, added to fugue of 16 June 1827; R xi/3	
54	Variations sérieuses, d, 4 June 1841 (Vienna, 1843); R xi/2	266
82	Variations, E♭, 25 July 1841 (1849); R xi/2	
83	Variations, B♭, ?1841 (1850); R xi/2	
—	Andante, E♭, c June 1842, orig. part of op.72, ed. H. O. Hieckel (1969)	
72	Sostenuto, F, c June 1842, orig. part of op.72, *GB-Ob	266
—	Kinderstücke ('Christmas Pieces') (London, 1847); G, 24 June 1842, E♭, G, 21 June 1842, D, g, F; R xi/2	
—	Lied [ohne Worte], D, c19 Jan 1843, inc., *S-Smf	
—	Lied ohne Worte, D, 18 March 1843, *D-Bds	
62	Lieder ohne Worte, v (Bonn, 1844): G, 12 Jan 1844, B♭, 29 July 1843, 'Trauermarsch', e, 19 Jan 1843, G, Venezianisches Gondellied, a, 'Frühlingslied', A, 1 June 1842, R xi/4	266
—	Lied ohne Worte ('Allegro marcato alla marcia'), d, 12 Dec 1844, ed. E. Walker as Reiterlied (1947)	

67   Lieder ohne Worte, vi (Bonn, 1845): Eb, 29 July 1843, f#, 3 May 1845, Bb, May 1845, Bb, 23 Nov ?1844, 'Spinnerlied', C, 5 May 1845, b, 5 Jan 1844, E, 29 April, 1841; R xi/4 — 266

85   Lieder ohne Worte, vii (Bonn, 1850): F, 29 April 1841, a, 9 June 1834, Eb, 19 Aug 1835, D, 6 May 1845, A, 7 May 1845, Bb, 1 May 1841; R xi/4 — 266

102 Lieder ohne Worte, viii (Bonn, 1868): e, 1 June 1842, D, 11 May 1845, C, 12 Dec 1845, g, ? 4 Feb 1841, Kinderstück, A, 12 Dec 1845, C; R xi/4 — 266

—   Two Piano Pieces, Bb, g (1860); R xi/3

—   Con moto, bb, *Bds

—   Study, F, *Bds

—   Two Piano Pieces: Andante, E, Allegro, e: *Bds

—   Lied ohne Worte ('Andante tranquillo'), D, *Bds

—   Lied [ohne Worte], Eb, inc., *Ob

—   Sonata, Eb, inc., *Ob

—   Fugue, Eb, inc., *Ob

—   Andante sostenuto, E, inc., *Ob

—   Allegretto, a, ded. M. Benecke, lost

—   Lied ohne Worte, F, ded. Doris Loewe, see Kahn (1924)

PIANO DUET

—   Lento–Vivace, g, 1820, *D-Bds

—   Fantasia, d, 15 March 1824, *Bds

92   Allegro brillant, A, 23 March 1841 (1851); R x

83a Variations, Bb, 10 Feb 1844 (1850), based on op.83 for pf solo; R x

—   Seven Lieder ohne Worte, arr. of op.62 nos.1–6 and op.67 no.1, 9 June 1844, ded. Prince Albert

—   Andante, g, inc., *GB-Ob

TWO PIANOS

—   Sonata, D, ? before 1821, *Ob [attrib. Mendelssohn]

—   Sonata, g, 21 Feb ? before 1821, *Ob [attrib. Mendelssohn]

—   Duo concertant, variations on march from Weber's Preciosa, c April 1833, perf. London, 1 May 1833, collab. Moscheles (1834, later as Moscheles's op.87b), orig. with orch

ORGAN

—   Five little pieces, *D-Bds: Fugue, d, Fugue, g, both 3 Dec 1820, — 267

—   Fugue, d, 6 Jan 1821, ed. W. A. Little (Leipzig, 1977), untitled, d, Dec 1820, Prelude, d, 28 Nov 1820, ed. L. Altman (1969) — 266

—   Etude für Orgel ('Nachtszene'), ?1821–2

—   Fantasia and Fugue, g, ?1822–3, *Bds (inc.)

—   Wie gross ist des Allmäch'gen Güte, chorale-prelude, 3 variants, 30 July–2 Aug 1835, ed. W. A. Little (1977)

—   Two little pieces, *Bds: Andante, D, 9 May 1823, ed. W. A. Little (1977), untitled [Passacaglia], c, 10 May 1823

—   Organ piece, A, for wedding of Fanny Mendelssohn, 3 Oct 1829, lost, ? reused in op.65 no.3

—   Nachspiel, D, 8 March 1831, ed. W. A. Little (1977), reused in op.65 no.2

—   Two Fugues for the Organ, 11 Jan 1835, ded. T. Attwood [duet arr. of op.37 no.1 and op.35 no.2, *B]

37   Three Preludes and Fugues (1837): c, prelude 2 April 1837, fugue 30 July 1834; G, prelude 4 April 1837, fugue 1 Dec 1837; d, prelude 6 April 1837, fugue 29 March 1833; early versions of fugues 1 and 3, ed. W. A. Little (1977); R xii — 267

—   Fugue, e, 13 July 1839, ed. L. Altman (1962)

—   Fugue, C, 14 July 1839, reused in op.65 no.2

—   Fugue, f, 18 July 1839 (London, 1885), perf. London, 30 Sept 1840, ed. L. Altman (1962)

—   O Haupt voll Blut und Wunden, chorale-prelude, ? c6 Aug 1840, inc., *GB-Ob

—   Prelude, c, 9 July 1841, pubd as suppl. to Scottish Musical Monthly (Dec 1894), ed. L. Altman (1969)

—   Three little pieces, *D-Bds: Allegro, F, 21 July 1844, Allegretto, d, 22 July 1844, Allegro, d, 25 July 1844

—   Two Pieces (London, 1898): Andante with variations, D, 23 July 1844, Allegro, Bb, 31 Dec 1844, ed. L. Altman (1966, 1969)

—   Chorale, Ab, 10 Sept 1844, *Bds

65   Six Sonatas (1845): f/F, 18 Dec 1844, c/C, 21 Dec 1844, A, 17 Aug 1844, Bb, 2 Jan 1845, D, 9 Sept 1844, d/D, 27 Jan 1845; R xii — 267

—   Andante alla marcia, Bb, 2 Jan 1845, reused in op.65 no.4

—   Andante sostenuto, D, 26 Jan 1845, reused in op.65 no.6

—   Fugue, Bb, 2 April 1845, reused in op.65 no.4

—   Chorale, D, *Bds

PSALMS, SACRED CANTATAS, LARGER SACRED WORKS		
—	Gloria, E♭, solo vv, chorus, orch, inc., c March 1822, *D-Bds*	250
—	Psalm lxvi, double female chorus, bc, 8 March 1822, *Bds*	204
—	Magnificat, D, solo vv, chorus, orch, 31 May 1822, *Bds*	204, 250
—	Salve regina, E♭, S, str, ?1824, pubd in R. Werner (1930)	
—	Two sacred pieces, chorus, *Bds*: Wie gross ist des Allmächt'gen Güte, ?1824, Allein Gott in der Höh' sey Ehr, 10 Sept 1824	
—	Kyrie, d, 5-part chorus, orch, 6 May 1825, *GB-Ob*, vocal score, ed. R. Leavis (1964)	
—	Te Deum, D, solo vv, double chorus, bc, 5 Dec 1826, perf. Berlin, 12 Feb 1827; L vi/1	250
—	Christe, du Lamm Gottes, chorale cantata, chorus, orch, Christmas 1827, ed. O. Bill (Stuttgart, 1978)	
—	Jesu, meine Freude, chorale cantata, chorus, str, 22 Jan 1828 facs. ed. O. Jonas (Chicago, 1966), ed. B. Pritchard (Hilversum, 1972)	
—	Wer nur den lieben Gott lässt walten, chorale cantata, solo v, chorus, str, c April–July 1829, ed. O. Bill (Kassel, 1976)	
—	O Haupt voll Blut und Wunden, chorale cantata, solo v, chorus, orch, 13 Sept 1830, ed. R. L. Todd (Madison, 1981)	
31	Psalm cxv, solo vv, chorus, 15 Nov 1830 (Bonn, 1835), orig. after Vulgate version, Non nobis Domine; R xiv/A1	
—	Vom Himmel hoch, chorale cantata, solo vv, chorus, orch, 28 Jan 1831, ed. K. Lehmann (Stuttgart, 1983)	
—	Verleih uns Frieden, chorus, orch, 10 Feb 1831; facs. in *AMZ*, xli, suppl. for 5 June 1839; R xiv/A3	
—	Wir glauben all an einen Gott, chorale cantata, chorus, orch, c March 1831, ed. G. Graulich (Stuttgart, 1980)	
—	Ach Gott vom Himmel sieh' darein, chorale cantata, solo vv, chorus, orch, 5 April 1832, ed. B. Pritchard (1972)	
—	Te Deum, Morning Service, A, solo, vv, chorus, org, c Aug 1832 (London, 1832); R xiv/B	
121	Responsorium et Hymnus, Vespergesang, male vv, vc, org, 5 Feb 1833 (1873); R xiv/B	
42	Psalm xliii, solo vv, chorus, orch, org, 1st version, c July 1837; final version, 22 Dec 1837 (1838); R xiv/A1	216
46	Psalm xcv, T, chorus, orch, 6 Aug 1838, rev. early 1839 and 3 July 1841 (1842); R xiv/A1	
—	Psalm v 'Lord hear the voice', chorus, 26 Feb 1839, *D-Bds*	
—	Psalm xxxi 'Defend me, Lord', chorus, 27 Feb 1839, pubd in National Psalmodist (1840); *D-Bds*	250
51	Psalm cxiv, 8vv, orch, 9 Aug 1839 (1841); R xiv/A1	
—	Several psalm melodies and harmonizations, chorus, ? 13 Nov 1843	
91	Psalm xcviii, double chorus, orch, org, 27 Dec 1843, perf. Berlin, 1 Jan 1844 (c1851); R xiv/A1	227, 250
—	Psalm c, 'Jauchzet den Herrn', chorus, ? 1 Jan 1844, ed. in Musica sacra, viii; R xiv/C	250
78	Three psalms (1848): Ps ii, solo vv, chorus, 15 Dec 1843; Ps xliii, 8vv, 3 Jan 1844; Ps xxii, solo vv, chorus, 1844; R xiv/C	227, 250
73	Lauda Sion, solo vv, chorus, orch, 6 Feb 1846, perf. Liège, 11 June 1846 (1849); R xiv/A3	231, 250
—	Er wird öffnen die Augen der Blinden, chorus, orch, *Bds*, intended for Elijah (?1846)	
—	Kyrie, chorus, orch, *Bds*	
MOTETS, ANTHEMS, OTHER SHORTER SACRED PIECES		
—	Die Himmel erzählen (Ps xix), 5vv, 16 June 1820, *Bds*	
—	Gott, du bist unsre Zuversicht (Ps xix), 5vv, ?1820, *Bds*	
—	Ich will den Herrn nach seiner Gerechtigkeit preisen, 4vv, ?1820, *Bds*	
—	Tag für Tag sei Gott gepriesen, 5vv, ?1820, *Bds*	
—	Das Gesetz des Herrn ist ohne Wandel (Ps xix), 6vv, ?1821, *Bds*	
—	Deine Rede präge ich meinem Herzen (Ps cxix), 4vv, ?1821, *GB-Ob*	
—	Ein Tag sagt's dem andern (Ps xix), S, A, pf, ?1821, *D-Bds*	
—	Er hat der Sonne eine Hütte gemacht (Ps xix), 5vv, ?1821, perf. of Ps xix, Berlin, 18 Sept 1821, *Bds*	
—	Ich weiche nicht von deinen Rechten (Ps cxix), 4vv, ?1821, *GB-Ob*	
—	Jube Domine, solo vv, double chorus, 25 Oct 1822, ed. G. Graulich (Stuttgart, 1980)	
—	Kyrie, c, solo vv, double chorus, 12 Nov 1823, ed. G. Graulich (Stuttgart, 1980)	
—	Jesus, meine Zuversicht, solo vv, chorus, pf, 9 June 1824, *D-Bds*	
111	Tu es Petrus, 5vv, orch, 14 Nov 1827 (Bonn, c1868); R xiv/A3	

Ave maris stella, S, orch, 5 July 1828, *Bds

Hora est, 16vv, org, 6 Dec 1828, perf. Berlin, 14 Nov 1829, ed. M. Hutzel (Stuttgart, 1981)

23 Three sacred pieces (Bonn, 1832): Aus tiefer Noth, T, chorus, org, 19 Oct 1830; Ave Maria, solo vv, 8vv, bc, 16 Oct 1830; Mitten wir im Leben sind, 8vv, 20 Nov 1830; R xiv/B

Zum Feste der Dreieinigkeit ('O beata et benedicta'), 3 S, org, 30 Dec 1830, *Bds

39 Three Motets, female chorus, org, 31 Dec 1830, rev. 1837/8 (Bonn, 1838): Hear my prayer, O Lord ('Veni, Domine'), O praise the Lord ('Laudate pueri') 14 Aug 1837, O Lord, thou hast searched me out ('Surrexit Pastor'); R xiv/B

Lord have mercy upon us, To the Evening Service, chorus, 24 March 1833 pubd in Album für Gesang (1842); R xiv/C

115 Two sacred choruses, 'Beati mortui', 'Periti autem', male chorus, ?1833–4 (? c 1870); R xiv/C

96 Hymn (Ps xiii), A solo, chorus, orch, 5 Jan 1843 (Bonn, ? c1852); first 3 movts with org acc., 12 Dec 1840 (Bonn, 1841); R xiv/A3; xiv/B

Herr Gott, dich loben wir, solo vv, double chorus, orch, org, 16 July 1843, perf. Berlin, 6 Aug 1843, *Bds

Chorale harmonizations, chorus, wind, Dec 1843: Allein Gott in der Höh, Vom Himmel hoch, Wachet auf

Ehre sei dem Vater, 8vv, 17 Jan 1844, *Bds

Hear my prayer (paraphrase of Ps lv by W. Bartholomew), hymn, S, chorus, org, 25 Jan 1844, perf. London, 8 Jan 1845 (Berlin, 1845), R xiv/B; orch arr., Feb 1847

Denn er hat seinen Engeln befohlen (Ps xci), double chorus, 15 Aug 1844 (Berlin, n.d.), reused in Elijah, *Bds

Ehre sei dem Vater, C, 4vv, 5 March 1845, *Bds

Er kommt aus dem kindlichen Alter der Welt, 6vv, ?25 Jan 1846, *Bds

Cantique pour l'Eglise wallonne de Francfort ('Venez chantez'), 4vv, 1846, *Bds

79 Six Anthems, double chorus, Oct 1846 (1848): Rejoice, O ye people, 15 Dec 1843; Thou, Lord, our refuge hast been, 25 Dec 1843; Above all praises, 9 Oct 1846; Lord, on our

offences, 14 Feb 1844; Let our hearts be joyful, 5 Oct 1846; For our offences, 18 Feb 1844; R xiv/C

Die deutsche Liturgie, 8vv, 28 Oct 1846: Kyrie, Heilig, Ehre sei Gott in der Höhe, ed. in Musica sacra, vii (Berlin, 1838); R xiv/C; Und Friede auf Erden, Wir loben dich, Denn du allein bist heilig, Amen; all unpubd, *Bds

69 Three English Church Pieces, solo vv, chorus, 1847 (?1847): Nunc dimittis, 12 June 1847, Jubilate, 5 April 1847, Magnificat, 12 June 1847; R xiv/C

Gloria patri ('Ehre sei dem Vater'), 4vv, *Bds

Glory be to the Father, 4vv, *Bds

SECULAR CANTATAS

In feierlichen Tönen, wedding cantata, S, A, T, chorus, pf, 13 June 1820, *Bds — 247–8

Grosse Festmusik zum Dürerfest (K. Levetzow), solo vv, chorus, orch, 1828, perf. Berlin, 18 April 1828, *Bds — 208, 250

Begrüssung ('"Humboldt" Cantata') (L. Rellstab), festival music, solo male vv, male chorus, orch (with timp, vc and db), 12 Sept 1828, perf. Berlin, 18 Sept 1828 — 208, 249

60 Die erste Walpurgisnacht (Goethe), cantata, chorus, orch, 13 Feb 1832, perf. Berlin, 10 Jan 1833, rev. 1842–3, perf. Leipzig, 2 Feb 1843 (1844); R xv — 211, 213, 225, 229, 249–50, 251

Gott segne Sachsenland, male vv, wind, 2 June 1843, perf. Dresden, 7 June 1843, *Bds

68 An die Künstler (Schiller), festival song, male vv, brass, 19 April 1846, perf. Cologne, June 1846 (Bonn, 1846); R xv — 231, 250

CHORAL SONGS

Einst ins Schlaraffenland zogen, 4 male vv, 1820, *Bds — 250–51

Lieb und Hoffnung, male vv, 1820, *Bds

Quis sit desiderio, (Horace: Odes I/xxiv), chorus, ? c1820 *Cincinnati Art Museum

Jägerlied ('Kein bess're Lust in dieser Zeit') (L. Uhland), 4 male vv, 20 April 1822, *Bds

Lob des Weines ('Seht, Freunde, die Gläser'), solo male vv, male chorus, 1822, *Bds

Lasset heut am edlen Ort (Goethe), 4 male vv, 19 May 1828,

Der Sänger (Schiller), 30 Oct 1845, perf. Leipzig, 10 Nov 1846

75 Four Male Choruses, male vv (1849); R xvii
1 Der frohe Wandersmann (Eichendorff), 8 Feb 1844
2 Abendständchen (Eichendorff), 14 Nov 1839
3 Trinklied (Goethe)
4 Abschiedstafel (Eichendorff), ? 12 Feb 1838, *US-Cn (copy)

76 Four Male Choruses (1849); R xvii
1 Das Lied vom braven Mann (Heine)
2 Rheinweinlied (G. Herwegh), 8 Feb 1844
3 Lied für die Deutschen in Lyon (F. Stoltze), 8 Oct 1846
4 Comitat (Hoffmann von Fallersleben), 14 Sept 1847

88 Six Choruses, mixed vv (c1850); R xvi
1 Neujahrslied (J. P. Hebel), 8 Aug 1844
2 Der Glückliche (Eichendorff), 20 June 1843
3 Hirtenlied (Uhland), 14 June 1839
4 Die Waldvögelein (Schütz), 19 June 1843
5 Deutschland (E. Geibel), ?1839–43
6 Der wandernde Musikant (Eichendorff), 10 March 1840

100 Four Choruses, mixed vv (1852); R xvi
1 Andenken, 8 Aug 1844
2 Lob des Frühlings (Uhland), 20 June 1843
3 Frühlingslied, ?1843–4
4 Im Wald, 14 June 1839

120 Four Male Choruses (c1873); R xvii
1 Jagdlied (W. Scott), 27 Nov 1837
2 Morgengruss des Thüringischen Sängerbundes, 20 Feb 1847
3 Im Süden, 24 Nov 183~
4 Zigeunerlied (Goethe)

— Lob der Trunkenheit ('Trunken müssen wir alle sein') 4 male vv, *Bds

— Musikantenprügelei ('Seht doch diese Fiedlerbanden'), 4 male vv, pubd as suppl. to Die Musik, viii/2 (1908–9)

CONCERT ARIAS

— Che vuoi mio cor?, Mez, str, *Bds

— Ch'io t'abbandono (Metastasio: Achille in Sciro), Bar, pf, 5 Sept 1825, *US-NYpm

---

facs. in Festlied zu Zeeters siebzigsten Geburtstag (1928)

— Worauf kommt es überall an, 4 male vv, 23 Feb 1837, *Bds

41 Im Freien zu singen, mixed vv (1838); 2–4 as Drei Volkslieder, R xvi — 251
1 Im Walde (A. von Platen), Jan 1838
2 Entflieh mit mir (Heine), 22 May 1835
3 Es fiel ein Reif (Heine), 22 May 1835
4 Auf ihrem Grab (Heine), 22 Jan 1834
5 Mailied (L. Hölty), 22 May 1835
6 Auf dem See (Goethe), 22 May 1835

48 Der erste Frühlingstag, mixed vv (1840); R xvi
1 Frühlingsahnung (Uhland), 5 July 1839
2 Die Primel (N. Lenau), 1839
3 Frühlingsfeier (Uhland), 28 Dec 1839
4 Lerchengesang, canon, 15 June 1839
5 Morgengebet (Eichendorff), 18 Nov 1839
6 Herbstlied (Lenau), 26 Dec 1839

— Ersatz für Unbestand (F. Rückert), 4 male vv, 22 Nov 1839, pubd in Deutscher Musenalmanach [Dec 1839]; R xvii

— Festgesang [for the Gutenberg Festival] (A. E. Prölss), male vv, 1840, perf. Leipzig, 25 June 1840 (n.d.); R xv [no.2 adapted by W. H. Cummings as 'Hark! the Herald Angels Sing']

50 Six Male Choruses (?1840); R xvii
1 Türkisches Schenkenlied (Goethe), ?1839–40
2 Der Jäger Abschied (Eichendorff), 4 hn, b trbn acc., 6 Jan 1840
3 Sommerlied (Goethe), 1839–40
4 Wasserfahrt (Heine), ?1839–40
5 Liebe und Wein, 7 Dec 1839
6 Wanderlied (Eichendorff), 6 Jan 1840

— Nachtgesang, 4 male vv, 15 Jan 1842 (n.d.); R xvii — 251
— Die Stiftungsfeier, 4 male vv, 15 Jan 1842 (n.d.); R xvi — 251

59 Im Grünen, mixed vv (1844); R xvi — 251
1 Im Grünen (H. von Chezy), 23 Nov 1837
2 Frühzeitiger Frühling (Goethe), 17 June 1843
3 Abschied vom Wald (Eichendorff), 4 March 1843
4 Die Nachtigall (Goethe), 19 June 1843
5 Ruhetal (Uhland), 4 March 1843
6 Jagdlied (Eichendorff), 5 March 1843 — 251

116 Sahst du ihn herniederschweben, funeral song, mixed vv, 8 July

Des Mädchens Klage (Schiller) (n.d.); R xix

— Gretchen ('Meine Ruh ist hin') (Goethe), *D-Bds (inc.)

— Es rauscht der Wald, *Bds

— Vier trübe Monden sind entfloh'n (Munich, 1882)

— Weinend seh' ich in die Nacht, *Bds

— An Marie ('Weiter, rastlos atemlos vorüber'), (Munich, 1882)

— Erwartung ('Bist auf ewig du gegangen'), frag. (Munich, 1882)

— Ja, war's nicht aber Frühlingszeit (Hoffmann von Fallersleben), Charlotte und Werther, lost

VOCAL DUETS
(all with pf acc.)

63 Six Duets (1845); R xviii
1 Ich wollt' meine Lieb' (Heine), 3 Dec 1836
2 Abschied der Zugvögel (Hoffmann von Fallersleben), 20 May 1844
3 Gruss (Eichendorff), March 1844
4 Herbstlied (Klingemann), 16 Oct 1836
5 Volkslied (R. Burns), 17 Oct 1842
6 Maiglöckchen und die Blümelein (Hoffmann von Fallersleben), 23 Jan 1844

77 Three Duets (1848); R xviii
1 Sonntagsmorgen (Uhland), 3 Dec 1836
2 Das Aehrenfeld (Hoffmann von Fallersleben), 18 Jan 1847
3 Lied aus 'Ruy Blas' (Hugo), 14 Feb 1839, pubd in A. Schmidt, ed.: Orpheus, musikalisches Taschenbuch für das Jahr 1840 (Vienna and Leipzig, 1840)

— Three Folksongs (Berlin, 1847); R xviii
1 Wie kann ich froh und lustig sein (P. Kaufmann)
2 Abendlied (Heine), 19 Jan 1840
3 Wasserfahrt (Heine)

CANONS
(source unknown or in private collection unless otherwise stated)

— Canon on motif from Mozart's 'Jupiter' Symphony K551, 4 Nov [1821] [in letter to E. Rietz], *US-NYp

— Three-part canon, 13 May 1825, ded. S. Neukomm

Kurzgefasste Übersicht des canonischen Rechts, 3 vn, 6 Feb 1827, ded. H. Romberg

— Three-part canon, 27 Sept 1827, ded. F. Hiller, facs. in Jb des Kölnischen Geschichtsvereins, xli (1967)

— Three-part canon, 9 April 1829, ded. Henriette Sontag

— Three-part canon, 19 May 1830, ded. H. Dorn, *D-Zsch

— Two-part canon, 2 va, 26 June 1831, ded. G. Smart, *US-NYpm

— Three-part canon, 22 March 1832

— Four-part canon, b, 16 April 1832, ded. F. Chopin, facs. in L-Binental: Chopin: Dokumente und Erinnerungen aus seiner Heimstadt (Leipzig, 1932), pl. xxxix

— Four-part canon, 'Wohl ihm', 30 May 1832, ded. I. Moscheles

— Der weise Diogenes, 4 male vv, 11 Feb 1833, facs. suppl. to Die Musik, viii/2 (1908–9)

— Three-part canon, 11 April 1833, on text by G. Nauenburg, *F-Pc

— Three-part canon, 16 Dec 1835, ded. C. Künzel

— Und ob du mich züchtigst, 5vv, 24 Dec 1835, *D-Bds

— Three-part canon, 19 April 1836, *Hanover, Kestner-Museum

— Three-part canon, May 1836

— Four-part canon, 7 Sept 1837, *GB-Lbm

— Two-part canon, b, 9 Sept 1837, *D-Bds

— Three-part canon, 17 Sept 1837, ded. Charles Woolloton, *B

— Two-part canon, b, 24 Sept 1837, *GB-Ob

— Two-part canon, b, pf, 24 Sept 1837, *D-Bds

— Two-part canon, c, pf, Jan 1838, *Bds

— Two-part canon, 10 Feb 1839, *F-Pc

— Two-part canon, March 1839, ded. F. Whistling, *US-NYp

— Two-part canon, 12 April 1839, ded. Kietz [painter]

— Three-part canon, 7 Sept 1839, ded. B. Müller, *D-B [uses canon of 16 April 1832]

— Three-part canon, 8 Sept 1839, *Bds

— Four-part canon, 4 Dec 1839, ded. A. Hesse, *F-Pn

— Canon, 11 Dec 1839, ded. A. Heyse, *US-Wc

— Two-part canon, 14 Feb 1840, *D-Bds [solution by F. Möhring]

— Two-part canon, 11 Nov 1840, ded. H. C. Andersen, *DK-KK

— Two-part canon, 6 Jan 1841, *US-Wc [uses canon of 9 Sept 1837]

Canon, E♭, 28 March 1841

— Two-part canon, 7 April 1841, ded. R. Lepsius, *D-Bds

— Two-part canon, 22 April 1841

— Two-part canon, pf, 22 April 1841, ded. V. Carus, *Bds

— Canone a 3, 'Pater peccavi', 7 Aug 1841, ded. F. Kistner, *US-Wc

— Two-part canon, 23 Dec 1841, ded. Leon Herz, *D-B (copy)

— Canon, f♯, 7 June 1842, *GB-Ob

— Canone a 3 unis., 11 July 1842, ded. Felix Moscheles [uses canon of 17 Sept 1837; facs. in Letters of Felix Mendelssohn to Ignaz and Charlotte Moscheles (Boston, 1883), 225]

— Canon, 7 July 1843

— Etude, vn, or canon, 2 vn, 11 March 1844, ded. J. Joachim, facs. in O. E. Deutsch, ed.: St Cecilia's Album (Cambridge, 1944)

— Three-part canon, 8 July 1844, facs. in G. Kinsky, ed.: Musikhistorisches Museum von Wilhelm Heyer in Cöln: Katalog (Leipzig, 1910), i, 339

— Scherzo osia canone a 4, 8 July 1844

— Canon, 9 July 1844, *US-AUS

— Canon, f♯, 2 Nov 1844, ded. L. Lallemant, facs. in H.-J. Rothe and R. Szeskus (1972)

— Canon, A♭, 4 vv, 12 July 1845, ded. Frau A. von Woringen, facs. in Wolff (1909), 168

— Canon, 5 Feb 1846, *S7u [uses canon of 9 Sept 1837]

— Ratselkanon, 19 Feb 1846

— Two-part canon, 18 April 1846, ded. M. Szymanowska, facs. in M. Szymanowska: Album (Kraków, 1953)

— Three-part canon, 20 April 1846, *Milan, Museo della Scala

— Two-part canon, 30 June 1846, *F-Pc

— Canone doppio, 29 July 1846, ded. J. Warburg

— Two-part canon, 26 Aug 1846, facs. in The Autographic Mirror (London, 1864), ii

— Two-part canon, 2 Sept 1846, ded. A. de Chene de Vere, *US-Wc

— Canon, f♯, 6 May 1847, *GB-Lbm [uses canon of 7 June 1842]

— Four-part canon, c, subject from opening of Beethoven's Piano Concerto no.3 op.37, *D-Bds

— Four-part double canon, *Bds

— Two-part canon, b, *GB-Ob

See also 'Choral songs', op.48 no.4

COMPOSITION EXERCISES

Workbook with exercises in figured bass, chorale, invertible counterpoint, canon and fugue a2, a3, c1819-21, ed. R. L. Todd (1983) [also incl. Movement, g, vn, pf; Theme and Variations, C, vn, pf; Andante, d, vn, pf; Allegro, C, vn, pf; Theme and Variations, D, pf; Four little pieces, pf; G, g, G, g; Pauvre Jeanette, 1v, pf; see 'Chamber', 'Piano Solo', 'Solo Songs']

TRANSCRIPTIONS AND ARRANGEMENTS

Bach

Organ Compositions on Chorales, i–iv (London and Leipzig, 1845–6) — 213, 214

44 kleine Choralvorspiele für die Orgel (Leipzig and London, 1845) — 214

15 grosse Choral-Vorspiele für die Orgel (Leipzig and London, 1846)

6 Variations on the Chorale 'Christ, der du bist der helle Tag – Christ who art the brightest day' (Leipzig and London, 1846) — 214

11 Variations on the Chorale 'Sey gegrüsset Jesu gütig – All hail good Jesus' (Leipzig and London, 1846)

Chaconne, vn, pf [London and Hamburg, 1847; Fr. edn, Paris and elsewhere, 1848] [pf acc. for the Chaconne in D minor for vn solo]

Suite, D, orch, ed. F. David (1866) [Mendelssohn's performing edn. for the Gewandhaus concerts]

Handel

Dettingen Te Deum, perf. Berlin, 1828 (1869)

Acis and Galatea, 3 Jan 1829, *GB-Ob (London, n.d.)

org parts for Solomon, 1834, and two choruses from Messiah, all unpubd, *D-Bds

Zadok the Priest, coronation anthem, wind pts., perf. Leipzig, 1 Jan 1836, *GB-Ob

Israel in Egypt (London, 1846) [Handel Society edn.] — 213

Joshua, org pt.

other

6 schottische National-Lieder, Ger. and orig. Eng. texts (1839); ed. R. Elvers (1977)

Beethoven: Marcia funebre, from Piano Sonata op.26, orch., *F-Pc

Cherubini: Ov. to Die Wasserträger, arr. pf duet, 9 Jan 1837, *GB-Ob

Cimarosa; Terzet, Il matrimonio segreto (?), arr. for Gewandhaus concert, 25 Feb 1847

J. Lang; Trinklied vor der Schlacht, arr. male chorus, ww, brass

Moscheles: Septet op.88, arr. pf duet, Aug 1833

Mozart: Ov. to Le nozze di Figaro, arr. pf duet, 14 Nov 1817 [?]

——: Piano Concerto K365/316a, cadenza, 1 June 1832, *LEbc

WRITINGS

Paphlëis, ein Spott-Heldengedicht, 1820, ed. M. F. Schneider (1961)

Trans. of Horace: Ars poetica, ii, 15 Oct 1820, *D-B

Das Mädchen von Andros, eine Komödie des Terentius, in den Versmassen des Originals übersetzt (Leipzig, 1826) [trans. of Terence: The Woman of Andros]

Draft of lib for oratorio, Moses, by A. B. Marx, Jan 1833

208

BIBLIOGRAPHY

CATALOGUES OF WORKS

'Felix Mendelssohn-Bartholdy', *AMZ*, xxxix (1837), 845

Vollständiges Verzeichnis im Druck erschienener Compositionen von . . . F. Mendelssohn-Bartholdy (Leipzig, 1841)

A. J. Becher: 'Vollständiges Verzeichniss der Compositionen von Dr. F. Mendelssohn-Bartholdy', *Orpheus-Almanach,* iii (Vienna, 1842), pp.iii–viii

Thematisches Verzeichnis der im Druck erschienenen Compositionen von Felix Mendelssohn Bartholdy (Leipzig, 1846; new edn., 1853, 2/1873, 3/1882/R1966 and 1973)

'Systematisches Verzeichnis der in Deutschland im Druck erschienenen Compositionen von Felix Mendelssohn Bartholdy', *Musikalisches Wochenblatt,* i (1870), suppl.

COLLECTIONS AND EXHIBITIONS

G. Kinsky, ed.: *Musikhistorisches Museum von Wilhelm Heyer in Cöln: Katalog* (Cologne, 1916), iv, 327

——: *Erstlingsdrucke der deutschen Tonmeister der Klassik und Romantik* (Vienna, 1934), 16 [on opp.1–3]

Y. Rokseth: 'Manuscrits de Mendelssohn à la Bibliothèque du Conservatoire', *RdM,* xv (1934), 103

E. Walker: 'An Oxford Collection of Mendelssohniana', *ML,* xix (1938), 426

M. F. Schneider: 'Eine Mendelssohn-Sammlung in Basel', *Der Amerbach Bote: Almanach* (Basle, 1947), 200

G. Kinsky: *Die Musikautographen-Sammlung L. Koch: Manuscripte, Briefe: Dokumente von Scarlatti bis Stravinsky* (Stuttgart, 1953)

E. Werner: 'Mendelssohn Sources', *Notes,* xii (1954–5), 201

Mendelssohn-Festwoche aus Anlass der 150. Wiederkehr des Geburtstages am 3. Februar 1959 (Leipzig, 1959)

R. Elvers: 'Verzeichnis der von Felix Mendelssohn Bartholdy herausgegebenen Werke J. S. Bachs', *Gestalt und Glaube: Festschrift für Oskar Söhngen* (Witten and Berlin, 1960), 145

M. F. Schneider: *Mendelssohn-Archiv der Staatsbibliothek Stiftung Preussischer Kulturbesitz* (Berlin, 1965)

——: *Die Wach'sche Mendelssohn-Sammlung auf dem Ried in Wilderswil bei Interlaken* (Berlin, 1966)

R. Elvers: 'Neuerwerbungen für das Mendelssohn-Archiv der Staatsbibliothek 1965–69', *Jb. der Stiftung Preussischer Kulturbesitz 1969,* 308

E. F. Flindell: 'Ursprung und Geschichte der Sammlung Wittgenstein im 19. Jahrhundert', *Mf,* xxii (1969), 300 [with list of Mendelssohn's letters]

Bibliography

M. Crum, ed.: *Felix Mendelssohn Bartholdy* (Oxford, 1972)

R. Elvers: *Felix Mendelssohn Bartholdy: Dokumente seines Lebens: Ausstellung zum 125. Todestag,* Staatsbibliothek Preussischer Kulturbesitz, Ausstellungskataloge, iii (Berlin, 1972)

P. Krause: *Autographen, Erstausgaben und Frühdrucke der Werke von Felix Mendelssohn Bartholdy in Leipziger Bibliotheken und Archiven* (Leipzig, 1972)

B. Richter: 'Das "Mendelssohn-Zimmer" in Leipzig', *Musik und Gesellschaft*, xxii (1972), 646

M. Crum: *Catalogue of the Mendelssohn Papers in the Bodleian Library, Oxford* (Tutzing, 1980, 1983)

J. R. Turner: 'Nineteenth-century Autograph Music Manuscripts in the Pierpont Morgan Library: Check List (II)', *19th Century Music*, iv (1980–81), 157

K. Schultz, ed.: *Felix Mendelssohn Bartholdy: 'Der schöne Zwischenfall in der deutschen Musik'* (Vienna, 1981)

ICONOGRAPHIES

M. F. Schneider: *Felix Mendelssohn im Bildnis* (Basle, 1953)

——: *Ein unbekanntes Mendelssohn-Bildnis von Johann Peter Lyser* (Basle, 1958)

LETTERS

F. Mendelssohn: *Reisebriefe aus den Jahren 1830 bis 1832*, ed. P. Mendelssohn Bartholdy (Leipzig, 1861, 8/1869; rev., enlarged P. Hübner, Bonn, 1947; Eng. trans., 1862, 9/1887/*R*1970) [vol.i of *Briefe aus den Jahren 1830 bis 1847*, compiled J. Rietz (Leipzig, 1861–3, 8/1915)]

——: *Briefe aus den Jahren 1833 bis 1847*, ed. P. and C. Mendelssohn Bartholdy (Leipzig, 1863; Eng. trans., 1863, 4/1864/*R*1970) [vol.ii of *Briefe aus den Jahren 1830 bis 1847*]

Pis'ma Mendel'sona-Bartol'di [Mendelssohn's letters] (St Petersburg, 1863)

L. Nohl: *Musiker-Briefe* (Leipzig, 1867; Eng. trans., 1867), 297–346 [30 letters of Mendelssohn]

F. Mendelssohn: *Acht Briefe* (Leipzig, 1871; Eng. trans. in *Macmillan Magazine*, London, June 1871) [letters to Henriette Voigt]

F. Hiller: *Felix Mendelssohn Bartholdy: Briefe und Erinnerungen* (Cologne, 1874, 2/1878; Eng. trans., 1874) [see also R. Sietz: *Aus Ferdinand Hillers Briefwechsel* (Cologne, 1958), 31ff]

E. Hanslick, ed.: 'Briefe von Felix Mendelssohn-Bartholdy an Aloys Fuchs', *Deutsche Rundschau*, lvii (1888), 65

F. Mendelssohn: *Briefe an Ignaz und Charlotte Moscheles*, ed. F. Moscheles (Leipzig, 1888/*R*1976; Eng. trans., 1888/*R*1970)

285

'Zwei Jugendbriefe Felix Mendelssohn-Bartholdy's', *Deutsche Rundschau*, liv (1888), 141

M. Friedländer: 'Ein Brief Felix Mendelssohns', *VMw*, v (1889), 483 [letter to F. von Piatkowski, 26 June 1838]

M. Freidländer, ed.: 'Briefe an Goethe von Felix Mendelssohn-Bartholdy', *Goethe-Jb*, xii (1891), 77, 110

J. Schubring, ed.: *Briefwechsel zwischen Felix Mendelssohn Bartholdy und Julius Schubring: Zugleich ein Beitrag zur Geschichte und Theorie des Oratoriums* (Leipzig, 1892/*R*1973)

E. Hanslick: 'Briefe von F. Mendelssohn', *Aus neuer und neuester Zeit* [*Die moderne Oper*, ix] (Berlin, 1900/*R*1971), 281 [descriptions of letters to F. Hauser]

E. Schirmer, ed.: 'Briefe Felix Mendelssohns an J. W. Schirmer', *Die Musik*, ii (1902–3), 83

W. Altmann: 'Zwei Briefe Mendelssohns', *Die Musik*, iv (1904–5), 179

H. B. and C. L. E. Cox: *Leaves from the Journals of Sir George Smart* (London, 1907/*R*1971) [incl. letters from Mendelssohn]

J. R. Sterndale Bennett: *The Life of William Sterndale Bennett* (Cambridge, 1907) [incl. correspondence with Mendelssohn]

E. Wolff, ed.: *Meister-Briefe: Felix Mendelssohn Bartholdy* (Berlin, 1907)

R. Scheumann, ed.: 'Briefe berühmter Komponisten aus dem Archiv des Königlichen Hof- und Domchores zu Berlin', *Die Musik*, viii (1908–9), 259

E. Wolff: 'Ein unveröffentlichter Brief von Felix Mendelssohn Bartholdy an Sigismond Neukomm', *Die Musik*, viii (1908–9), 338

——: 'Sechs unveröffentlichte Briefe Felix Mendelssohn Bartholdys an Wilhelm Taubert', *Die Musik*, viii (1908–9), 165

K. Klingemann [jr], ed.: *Felix Mendelssohn-Bartholdys Briefwechsel mit Legationsrat Karl Klingemann in London* (Essen, 1909)

M. Unger: *Von Mendelssohn-Bartholdys Beziehungen zu England: fünf englische Briefe des Meisters* (Langensalza, 1909)

E. Wolff: 'Briefe von Felix Mendelssohn-Bartholdy an seine rheinischen Freunde', *Rheinische Musik- und Theater-Zeitung*, x (1909), 86, 104, 121, 136, 149, 163, 182

'Unpublished Letters of Mendelssohn', *MT*, li (1910), 366 [letters to V. Novello, G. Hogarth, E. Buxton]

L. Dahlgren, ed.: *Bref till Adolf Fredrik Lindblad från Mendelssohn, . . . och andra* (Stockholm, 1913)

A. Mendelssohn Bartholdy: 'Felix Mendelssohn Bartholdy und Richard Wagner', *Programmbuch des 1. Fränkischen Musikfestes zu Würzburg* (Feb 1914) [incl. 4 letters from Wagner to Mendelssohn]

G. Fischer: 'Ein Brief des 15-jährigen Felix Mendelssohn Bartholdy',

Bibliography

Kleine Blätter (Hanover, 1916), 9 [letter to Friedrich Voigts]

H. Thompson: 'Some Mendelssohn Letters', *MT*, lxiv (1923), 461, 605 [letters to G. Macfarren]

G. Kleibömer: 'Ein ungedruckter Brief Felix Mendelssohns an Heinrich Romberg', *Die Musik*, xvi (1923–4), 826 [letter of 26 Jan 1844]

J. Tiersot, ed.: *Lettres de musiciens écrites en français du XVe au XIXe siècle (de 1831 à 1885)* (Paris, 1924)

R. Hübner: *Johann Gustav Droysen: Briefwechsel*, i (Berlin, 1929) [see also Wehmer (1959) below]

J.-W. Schottländer: 'Zelter und die Komponisten seiner Zeit', *Jb der Sammlung Kippenberg*, viii (1930), 238 [letter of 9 Aug 1822]

R. B. Gotch, ed.: *Mendelssohn and his Friends in Kensington: Letters from Fanny and Sophy Horsley Written 1833–36* (London, 1934)

'Mendelssohn, Sterndale Bennett and the Reid Professorship, an Unpublished Letter', *MT*, lxxxiv (1943), 351 [letter of 24 May 1844]

F. Mendelssohn: *Letters*, trans. G. Selden-Goth (New York, 1945/*R*1969)

H. Weiss: *Felix Mendelssohn-Bartholdy: ein Lebensbild in Briefen und zeitgenössischen Urteilen* (Berlin, 1947)

P. Sutermeister: *Felix Mendelssohn Bartholdy: Lebensbild mit Vorgeschichte: Reisebriefe von 1830–31* (Zurich, 1949)

O. E. Deutsch: The Discovery of Schubert's Great C-major Symphony: a Story in Fifteen Letters', *MQ*, xxxviii (1952), 528

H. Erdmann and H. Rentzow: 'Mendelssohns Oratorien-Praxis: ein bisher unbekannter Brief des Meisters vom Jahre 1841', *Musica*, vi (1952), 352 [letter of 22 March 1840 to Julius Stock]

H. E. Wilk: 'An Unpublished Letter by Felix Mendelssohn', *Musical America*, lxxii/3 (1952), 110 [letter to H. Grabau]

——: 'An Unpublished Mendelssohn Letter', *Canadian Musical Journal*, ii/2 (1958), 25 [letter to F. Kistner, 17 June 1839]

W. Reich: 'Mendelssohn sucht einen Operntext: fünf unbekannte Briefe des Komponisten', *Musica*, xiii (1959), 366 [letters to C. Birch-Pfeiffer]

C. Wehmer, ed.: *Ein tief gegründet Herz: der Briefwechsel Felix Mendelssohn-Bartholdys mit Johann Gustav Droysen* (Heidelberg, 1959)

E. Werner: 'The Family Letters of Felix Mendelssohn Bartholdy', *Bulletin of the New York Public Library*, lxv (1960), 5

F. Krautwurst: 'Briefe von Christian Heinrich Rinck, Felix Mendelssohn Bartholdy und Robert Schumann aus dem Nachlass J. G. Herzogs in der Erlangen Universitätsbibliothek', *Jb für Frankische Landesforschung*, xxi (1961), 149

R. Elvers: 'Ein Jugendbrief von Felix Mendelssohn', *Festschrift für Friedrich Smend* (Berlin, 1963), 95 [letter to Rudolph Gugel]

H. E. Samuel: 'Letters of Three Nineteenth Century Composers', *Cornell University Music Review*, vi (1963), 6 [letter of 26 Oct 1844]

'Ein unbekannter Brief Felix Mendelssohn Bartholdys', *BMw*, v (1963), 69 [letter to N. Simrock]

D. Schmidt: 'Felix Mendelssohn Bartholdy, ein Helfer der Abgebrannten von 1842', *Hamburgische Geschichts- und Heimatblätter*, viii (1967), 30

F. Mendelssohn Bartholdy: *Briefe*, ed. R. Elvers (Berlin, 1968–) [letters to German publishers]

Briefe von Felix Mendelssohn Bartholdy, 1833–1847 (Leipzig, 1968) [facs. edn. of 10 letters from *D-LEsm*]

G. Schulz: *Glückliche Jugend: Briefe des jungen Komponisten Felix Mendelssohn Bartholdy* (Bremen, 1971)

W. Anacker: 'Zwei Briefe von Felix Mendelssohn Bartholdy', *Musik und Gesellschaft*, xxii (1972), 654 [letters to A. F. Anacker]

H.-J. Rothe and R. Szeskus, eds.: *Felix Mendelssohn Bartholdy: Briefe aus Leipziger Archiven* (Leipzig, 1972)

F. Gilbert, ed.: *Bankiers, Künstler und Gelehrte: unveröffentlichte Briefe der Familie Mendelssohn aus dem 19. Jahrhundert* (Tübingen, 1975)

P. Krause: 'Ein unbekannter Brief von Mendelssohn', *Musik und Gesellschaft*, xxvi (1976), 429 [letter of 3 Feb 1840 to C. W. A. Porsche]

D. A. Wells: 'Letters of Mendelssohn, Schumann and Berlioz in Belfast', *ML*, lx (1979), 180 [letter to H. [?] Franck, 1 Jan 1840]

R. Elvers, ed.: *Mendelssohn Bartholdy: Briefe* (Frankfurt am Main, 1984)

FACSIMILE EDITIONS OF DOCUMENTS

Festlied zu Zelters siebzigsten Geburtstag MDCCCXXVIII, gedichtet von Goethe, vertont von Mendelssohn (Leipzig, 1928) [*Lasset heut am edlen Ort*]

F. Mendelssohn: *'Die Hebriden'*, ed. H. von Mendelssohn Bartholdy (Basle, 1947) [autograph of 2nd version]

M. F. Schneider, ed.: *Felix Mendelssohn Bartholdy: Denkmal in Wort und Bild* (Basle, 1947) [autographs of opp.30, 48/4, 88/3 and 100/4]

F. Mendelssohn: *Reisebilder aus der Schweiz, 1842*, ed. M. F. Schneider (Basle, 1954)

K. H. Köhler, ed.: *Die Frauen und die Sänger nach dem Gedicht 'Die vier Weltalter' von Friedrich Schiller für gemischten Chor komponiert* (Basle, 1959) [*Der Sänger*]

M. F. Schneider, ed.: *Lied einer Freundin: Zarter Blumen leicht Gewinde, ein bisher ungedrucktes Goethe-Lied von Mendelssohn* (Düsseldorf, 1960)

288

Bibliography

———: *Paphlëis, ein Spott-Heldengedicht* (Basle, 1961)

F. Mendelssohn: *Jesu meine Freude*, ed. O. Jonas (Chicago, 1966)

———: *Aquarellenalbum*, ed. M. F. Schneider and C. Hensel (Basle, 1968) [13 watercolour sketches in *D-B*]

R. Elvers, ed.: *Endreim-Spiele mit Felix Mendelssohn, Weimar, November 1821* (Berlin, 1970)

F. Mendelssohn: *Octet for Strings, Op.20*, ed. J. Newsom (Washington, DC, 1976) [autograph of 1st version]

MEMOIRS, RECOLLECTIONS

J. C. Lobe: 'Gespräche mit Felix Mendelssohn', *Fliegende Blätter für Musik* (Leipzig, 1855), i, 280

L. Rellstab: *Aus meinem Leben* (Berlin, 1861), ii, chap.11

H. Phillips: *Musical and Personal Recollections during Half a Century* (London, 1864), ii

A. B. Marx: *Erinnerungen: aus meinem Leben* (Berlin, 1865)

J. Schubring: 'Erinnerungen an Felix Mendelssohn Bartholdy', *Daheim*, ii (1866), 373; Eng. trans. in *Musical World*, xxxi (1866), 12 and 19 May

E. Polko: *Erinnerungen an Felix Mendelssohn-Bartholdy* (Leipzig, 1868; Eng. trans., 1869)

E. Devrient: *Meine Erinnerungen an Felix Mendelssohn-Bartholdy und seine Briefe an mich* (Leipzig, 1869, 3/1891; Eng. trans., 1869/*R*1972)

C. E. Horsley: 'Reminiscences of Mendelssohn', *Dwight's Journal of Music*, xxxii (1872), 345, 353, 361

E. Naumann: 'Erinnerungen an Felix Mendelssohn-Bartholdy', *Nachklänge: eine Sammlung von Vortragen und Gedenkblättern'*, (Berlin, 1872), 22

J. R. Planché: *Recollections and Reflections* (London, 1872, rev.1901), i, 279

C. Moscheles: *Aus Moscheles Leben* (Leipzig, 1872–3; Eng. trans., 1873) [incl. correspondence with Mendelssohn]

H. F. Chorley: *Autobiography, Memoir and Letters*, ed. H. G. Hewlett (London, 1873) [incl. correspondence with Mendelssohn]

J. Eckardt: *Ferdinand David und die Familie Mendelssohn-Bartholdy* (Leipzig, 1888)

H. S. Holland and W. S. Rockstro, eds.: *Memoir of Mme Jenny Lind-Goldschmidt* (London, 1891, abridged 2/1893) [incl. letters from Mendelssohn]

F. G. Edwards: 'Reminiscences of Mendelssohn', *MT*, xxxiii (1892), 465 [J. Benedict]

W. J. von Wasielewski: 'Felix Mendelssohn-Bartholdy und Robert Schumann: eine künstlerische Parallele mit Einflechtung persönlicher Erinnerungen', *Deutsche Revue*, xix/3 (1894), 329

289

Mendelssohn

H. Devrient, ed.: *Briefwechsel zwischen Eduard und Therese Devrient* (Stuttgart, 1909)
H. Davison: *Music during the Victorian Era: from Mendelssohn to Wagner* (London, 1912) [incl. letters of Mendelssohn to J. W. Davison]
E. von Webern: 'Felix Mendelssohn Bartholdy aus den Erinnerungen des Generalleutnants Karl Emil von Webern', *Die Musik*, xii (1912–13), 67
A. Mendelssohn Bartholdy: 'Erinnerungen an Felix Mendelssohn: aus alten Familienbriefen', *Neue freie Presse* (19 April 1925), suppl.
R. Schumann: *Erinnerungen an Felix Mendelssohn Bartholdy*, ed. G. Eismann (Zwickau, 1947, enlarged 2/1948; Eng. edn., 1951)

BIOGRAPHY, LIFE AND WORKS

W. A. Lampadius: *Felix Mendelssohn-Bartholdy: ein Denkmal für seine Freunde* (Leipzig, 1848; Eng. trans., 1876; Eng. trans. with suppl. sketches by J. Benedict, H. F. Chorley, L. Rellstab and others, ed. W. L. Gage, New York, 1865/*R*1978, 2/1866, and Boston, 1872, 2/1887)
L. Stierlin: *Biographie von Felix Mendelssohn-Bartholdy* (Zurich, 1849)
J. Benedict: *A Sketch of the Life and Works of the Late Felix Mendelssohn Bartholdy* (London, 1850, 2/1853)
V. Magnien: *Étude biographique sur Félix Mendelssohn-Bartholdy* (Beauvais, 1850)
W. H. Riehl: 'Bach und Mendelssohn aus dem socialen Gesichtspunkte', *Musikalische Charakterköpfe*, i (1853), 65–107
H. F. Chorley: 'The Last Days of Mendelssohn', *Modern German Music* (London, 1854/*R*1973), ii, 383–418
W. Neumann: *Felix Mendelssohn-Bartholdy: eine Biographie* (Kassel, 1854)
E. Krinitz [pseud. C. Selden]: *La musique en Allemagne: Mendelssohn* (Paris, 1867)
A. Reissmann: *Felix Mendelssohn-Bartholdy: sein Leben und seine Werke* (Berlin, 1867, rev., enlarged 3/1893)
H. Barbedette: *Felix Mendelssohn Bartholdy: sa vie et ses oeuvres* (Paris, 1868)
T. Marx: *Adolf Bernhard Marx' Vehältniss zu Felix Mendelssohn-Bartholdy* (Leipzig, 1869)
C. Mendelssohn Bartholdy: *Goethe und Felix Mendelssohn-Bartholdy* (Leipzig, 1871; Eng. trans. with addns by M. E. von Glehn, London, 1872, 2/1874/*R*1970)
H. Doering: *Leben und Wirken Felix Mendelssohn Bartholdys* (Wolfenbüttel, 1878)
S. Hensel: *Die Familie Mendelssohn 1729–1847, nach Briefen und*

290

Bibliography

Tagebüchern (Berlin, 1879, 18/1924; Eng. trans., 1881/*R*1969)

J. Sittard: *Felix Mendelssohn Bartholdy* (Leipzig, 1881)

A. Dörffel: *Geschichte der Gewandhausconcerte zu Leipzig vom 25. November 1781 bis 25. November 1881* (Leipzig, 1884/*R*1972), 83

W. S. Rockstro: *Mendelssohn* (London, 1884, rev., New York, 2/1911)

G. von Loeper: 'Mendelssohn-Bartholdy', *ADB*

W. A. Lampadius: *Felix Mendelssohn Bartholdy: ein Gesammtbild seines Lebens und Wirkens* (Leipzig, 1886/*R*1978)

J. C. Hadden: *Life of Mendelssohn* (London, 1888, 2/1904)

L. von Kretschman: 'Felix Mendelssohn-Bartholdy in Weimar: aus dem Nachlass der Baronin Jenny von Gustedt, geb. von Pappenheim', *Deutsche Rundschau*, lxix (1891), 304

S. S. Stratton: *Mendelssohn* (London, 1901, 6/1934)

G. Droysen: 'Johann Gustav Droysen und Felix Mendelssohn-Bartholdy', *Deutsche Rundschau*, cxi (1902), 107, 193, 386

E. van der Straeten: 'Mendelssohns und Schumanns Beziehungen zu J. H. Lübeck und Johann J. H. Verhulst', *Die Musik*, iii (1903–4), 8, 94

——: 'Streiflichter auf Mendelssohns und Schumanns Beziehungen zu zeitgenössischen Musikern', *Die Musik*, iv (1904–5), 25, 105

'Mendelssohn and his English Publisher' *MT*, xlvi (1905), 20, 167 [incl. letters to E. Buxton]

E. Wolff: *Felix Mendelssohn Bar.holdy* (Berlin, 1906, enlarged 2/1909)

C. Bellaigue: *Mendelssohn* (Paris, 1907, 4/1920)

P. de Stoecklin: *Mendelssohn* (Paris, 1908, 2/1927)

E. Rychnovsky: 'Aus Felix Mendelssohn Bartholdys letzten Lebenstagen', *Die Musik*, viii (1908–9), 141

B. Hake: 'Mendelssohn als Lehrer: mit bisher ungedruckten Briefen Mendelssohns an Wilhelm v. Boguslawski', *Deutsche Rundschau*, cxl (1909), 453

A. Mendelssohn Bartholdy: 'Felix Mendelssohn Bartholdy: Beitrag zur Geschichte seines Lebens und seiner Familie', *Frankfurter Zeitung* (31 Jan 1909), literary suppl.

W. Altmann: 'Mendelssohns Eintreten für Händel', *Die Musik*, xii (1912–13), 79

M. Jacobi: *Felix Mendelssohn Bartholdy* (Bielefeld, 1915)

W. Dahms: *Mendelssohn* (Berlin, 1919, 9/1922)

J. Esser: *Felix Mendelssohn Bartholdy und die Rheinlande* (diss., U. of Bonn, 1923)

W. H. Fischer: 'Felix Mendelssohn Bartholdy: sein Leben und Wirken in Düsseldorf', *Niederrheinisches Musikfest, Düsseldorf*, xcv (1926), 9–43

291

S. Kaufman: *Mendelssohn: 'a Second Elijah'* (New York, 1934, 2/1936/*R*1971)

J. Petitpierre: *Le mariage de Mendelssohn: 1837–1847* (Paris, 1937; Eng. trans., 1947, as *The Romance of the Mendelssohns*)

B. Bartels: *Mendelssohn-Bartholdy: Mensch und Werk* (Bremen, 1947)

J. Werner: 'Felix and Fanny Mendelssohn', *ML*, xxviii (1947), 303

K. H. Wörner: *Felix Mendelssohn-Bartholdy: Leben und Werk* (Leipzig, 1947)

G. Grove: *Beethoven–Schubert–Mendelssohn*, ed. E. Blom (London, 1951) [repr. from *Grove 1*]

N. van der Elst: 'Felix Mendelssohn en Cesar Franck', *Mens en Melodie*, viii (1953), 82

A. Ghislanzoni: 'I rapporti fra Spontini e Mendelssohn', *Congresso Intern. di Studi Spontiniani 1951: Atti* (Fabriano, 1954), 95

P. Radcliffe: *Mendelssohn* (London, 1954, rev. 2/1967)

R. Sterndale Bennett: 'The Death of Mendelssohn', *ML*, xxxvi (1955), 374

E. Werner: 'New Light on the Family of Felix Mendelssohn', *Hebrew Union College Annual*, xxvi (Cincinnati, 1955), 543

U. Galley: 'Bilder aus Düsseldorfs musikalischer Vergangenheit', *Neiderrheinisches Musikfest, Düsseldorf*, cx (1956), 33 [on Mendelssohn and K. Immermann]

H. C. Worbs: *Felix Mendelssohn Bartholdy* (Leipzig, 1956, 2/1957)

S. Jemnitz: *Felix Mendelssohn Bartholdy* (Budapest, 1958)

H. C. Worbs: *Felix Mendelssohn Bartholdy: Wesen und Wirken im Spiegel von Selbstzeugnissen und Berichten der Zeitgenossen* (Leipzig, 1958)

F. H. Franken: *Das Leben grosser Musiker im Spiegel der Medizin: Schubert, Chopin, Mendelssohn* (Stuttgart, 1959)

H. G. Reissner: 'Felix Mendelssohn-Bartholdy und Eduard Gans', *Publications of the Leo Baeck Institute*, iv (London, 1959), 92

R. Sietz: 'Mendelssohn ging nicht nach Weimar', *NZM*, Jg.120 (1959), 72

H. E. Jacob: *Felix Mendelssohn und seine Zeit: Bildnis und Schicksal eines Meisters* (Frankfurt am Main, 1959–60; Eng. trans., 1963)

E. Werner: 'Mendelssohn's Fame and Tragedy', *Reconstructionist*, xxv (1959–60), 9

M. F. Schneider: 'Mendelssohn und Schiller in Luzern', *Die Ernte*, xli (Basle, 1960), 125

———: *Mendelssohn oder Bartholdy? Zur Geschichte eines Familiennamens* (Basle, 1962)

R. Sietz: 'Das Stammbuch von Julius Rietz', *Studien zur Musikgeschichte des Rheinlandes*, lii (1962), 219

Bibliography

M. F. Schneider: 'Felix Mendelssohn Bartholdy: Herkommen und Jugendzeit in Berlin', *Jb der Stiftung Preussischer Kulturbesitz 1963*, 157

E. Werner: *Mendelssohn: a New Image of the Composer and his Age* (New York, 1963, rev. and enlarged, Zurich, 2/1980)

E. Rudolph: *Der junge Felix Mendelssohn: ein Beitrag zur Musikgeschichte der Stadt Berlin* (diss., Humboldt U., Berlin, 1964)

K. W. Niemöller: 'Felix Mendelssohn-Bartholdy und das Niederrheinische Musikfest 1835 in Köln', *Studien zur Musikgeschichte des Rheinlandes*, iii (Cologne, 1965), 46

K.-H. Köhler: *Felix Mendelssohn Bartholdy* (Leipzig, 1966, rev. 2/1972)

A. S. Kurtsman: *Mendel'son* (Moscow, 1967)

R. Sietz: 'Felix Mendelssohn und Ferdinand Hiller I: ihre persönlichen Beziehungen', *Jb des Kölnischen Geschichtsvereins*, xli (1967), 96

S. Grossmann-Vendrey: *Felix Mendelssohn Bartholdy und die Musik der Vergangenheit* (Regensburg, 1969)

M. Hurd: *Mendelssohn* (London, 1970)

W. Reich, ed.: *Felix Mendelssohn im Spiegel eigener Aussagen und zeitgenössischer Dokumente* (Zurich, 1970)

R. Sietz: 'Felix Mendelssohn und Ferdinand Hiller II: ihre künstlerischen Beziehungen', *Jb des Kölnischen Geschichtsvereins*, xliii (1971), 101

H. Kupferberg: *Felix Mendelssohn: his Life, his Family, his Music* (New York, 1972)

——: *The Mendelssohns: Three Generations of Genius* (New York, 1972)

G. R. Marek: *Gentle Genius: the Story of Felix Mendelssohn* (New York, 1972)

P. Ranft: *Felix Mendelssohn Bartholdy: eine Lebenschronik* (Leipzig, 1972)

Y. Tiénot: *Mendelssohn: musicien complet* (Paris, 1972)

B. Matthews: 'Mendelssohn and the Crosby Hall Organ', *MT*, cxiv (1973), 641

W. Blunt: *On Wings of Song: a Biography of Felix Mendelssohn* (New York, 1974)

H. C. Worbs: *Mendelssohn-Bartholdy* (Hamburg, 1974)

N. B. Reich: 'The Rudorff Collection', *Notes*, xxxi (1974–5), 247

E. Werner: 'Mendelssohniana dem Andenken Wilhelm Fischers', *Mf*, xxviii (1975), 19

H. Krellmann: 'Felix Mendelssohns Wirken im Rheinland', *Musica*, xxxi/6 (1977), 511

E. Werner: 'Mendelssohniana II, den Manen Egon Wellesz, des Freundes und Mentors', *Mf*, xxx (1977), 492

293

Mendelssohn

D. Jenkins and M. Visocchi: *Mendelssohn in Scotland* (London, 1978)
R. Grumbacher and A. Rosenthal: ' "Dieses einzige Stückchen Welt
 . . .": über ein Albumblatt von Felix Mendelssohn Bartholdy',
 *Totum me libris dedo: Festschrift zum 80. Geburtstag von Adolf
 Seebass* (Basle, 1979), 53
G. Schuhmacher: 'Wenn Komponisten erwachsen werden: zum
 Beispiel Mozart und Mendelssohn Bartholdy', *Melos/NZM*, Jg.
 144 (1983), 4

WORKS

O. Jahn: *Über Felix Mendelssohn-Bartholdys Oratorium Paulus: eine
 Gelegenheitsschrift* (Kiel, 1842)
——: 'Ueber Felix Mendelssohn Bartholdy's Oratorium *Elias*', *AMZ*,
 1 (1848), 113, 137
J. J. Bussinger: *Über Felix Mendelssohn und seine Musik zur Antigone*
 (Basle, c1862)
F. Zander: *Über Mendelssohns Walpurgisnacht* (Königsberg, 1862)
G. A. Macfarren: *Mendelssohn's Antigone* (London, 1865)
F. Chrysander: 'Mendelssohn's Orgelbegleitung zu Israel in Ägyp-
 ten', *Jahrbücher für musikalische Wissenschaft*, ii (Leipzig, 1867/
 *R*1966), 249
[? G. Grove]: 'Mendelssohn's Unpublished Symphonies', *MMR*, i
 (1871), 159
F. G. Edwards: 'Mendelssohn's "Hear My Prayer": a Comparison
 of the Original Ms. with the Published Score', *MT*, xxxii (1891),
 79
A. M. Little: *Mendelssohn's Music to the Antigone of Sophocles* (diss.,
 U. of London, 1893)
F. G. Edwards: 'Mendelssohn's Organ Sonatas', *PMA*, xxi (1894–5),
 1
——: *The History of Mendelssohn's Oratorio 'Elijah'* (London,
 1896/*R*1976)
J. W. G. Hathaway: *An Analysis of Mendelssohn's Organ Works: a
 Study of their Structural Features* (London, 1898/*R*1978)
F. G. Edwards: 'Mendelssohn's Organ Sonatas', *MT*, xlii (1901),
 794; xlvii (1906), 95
A. Heuss: 'Das "Dresdner Amen" im ersten Satz von Mendelssohns
 Reformationssinfonie', *Signale für die Musikalische Welt*, lxii
 (1904), 281, 305
O. A. Mansfield: *Organ Parts of Mendelssohn's Oratorios and other
 Choral Works Analytically Considered* (London, 1907)
A. Kopfermann: 'Zwei musikalische Scherze Felix Mendelssohns',

294

Bibliography

Die Musik, viii (1908–9), 179 [*Der weise Diogenes, Musikantenprügelei*]

O. A. Mansfield: 'Some Characteristics and Peculiarities of Mendelssohn's Organ Sonatas', *MQ*, iii (1917), 562

W. Kahl: 'Zu Mendelssohns Liedern ohne Worte', *ZMw*, iii (1920–21), 459

G. Schünemann: 'Mendelssohns Jugendopern', *ZMw*, v (1922–3), 506

J. Koffler: *Über orchestrale Koloristik in den symphonischen Werken von Felix Mendelssohn-Bartholdy* (diss., U. of Vienna, 1923)

J. Kahn: 'Ein unbekanntes "Lied ohne Worte" von Felix Mendelssohn', *Die Musik*, xvi (1923–4), 824 [*Lied ohne Worte*, F, ded. Doris Loewe]

H. Mandt: *Die Entwicklung des Romantischen in der Instrumentalmusik Felix Mendelssohn Bartholdys* (diss., U. of Cologne, 1927)

K. G. Fellerer: 'Mendelssohns Orgelstimmen zu Händelschen Werken', *HJb*, iv (1931), 79

G. Wilcke: *Tonalität und modulation im streichquartett Mendelssohns und Schumanns* (diss., U. of Rostock, 1932; Leipzig, 1933)

L. Hochdorf: *Mendelssohns 'Lieder ohne Worte' und der 'Lieder-ohne-Worte'-Stil in seinen übrigen Instrumentalwerken* (diss., U. of Vienna, 1938)

E. Walker: 'Mendelssohn's "Die einsame Insel",' *ML*, xxvi (1945), 148

A. M. Henderson: 'Mendelssohn's Unpublished Organ Works', *MT*, lxxxviii (1947), 347

H. and L. H. Tischler: 'Mendelssohn's Songs without Words', *MQ*, xxxiii (1947), 1

G. Abraham: 'The Scores of Mendelssohn's "Hebrides",' *MMR*, lxxviii (1948), 172

L. W. Leven: 'An Unpublished Mendelssohn Manuscript', *MT*, lxxxix (1948), 361 [canon, 7 April 1841]

A. van der Linden: 'Un fragment inédit du "Lauda Sion" de F. Mendelssohn', *AcM*, xxvi (1954), 48

E. Werner: 'Two Unpublished Mendelssohn Concertos', *ML*, xxxvi (1955), 126

J. Werner: 'The Mendelssohnian Cadence', *MT*, xcvii (1956), 17

D. Mintz: '*Melusine*: a Mendelssohn Draft', *MQ*, xliii (1957), 480

J. Werner: 'Mendelssohn's "Elijah", the 110th Anniversary', *MT*, xcviii (1957), 192

L. W. Leven: 'Mendelssohn's Unpublished Songs', *MMR*, lxxxviii (1958), 206

P. Losse: 'Ein bisher ungedrucktes Lied von Mendelssohn', *Musik und Gesellschaft*, ix (1959), 68 [*Lieben und Schweigen*]

Mendelssohn

K. Schönewolf: 'Mendelssohns Humboldt-Kantate', *Musik und Ge-sellschaft*, ix (1959), 408

H. C. Worbs: 'Die Entwürfe zu Mendelssohns Violinkonzert e-moll', *Mf*, xii (1959), 79

K.-H. Köhler: 'Zwei rekonstruierbare Singspiele von Felix Mendels-sohn Bartholdy', *BMw*, ii/3–4 (1960), 86

D. M. Mintz: *The Sketches and Drafts of Three of Felix Mendels-sohn's Major Works* (diss., Cornell U., 1960) [on *Elijah*, the 'Italian' Symphony and the D minor Trio]

A. Molnár: 'Die beiden Klavier-Trios in d-moll von Schumann (op.63) und Mendelssohn (op.49): eine Stiluntersuchung', *Sammel-bände der Robert-Schumann-Gesellschaft*, i (1961), 79

M. Rasmussen: 'The First Performance of Mendelssohn's *Festgesang* An die Künstler op.68', *BWQ*, iv (1961), 151

M. Geck: 'Sentiment und Sentimentalität im volkstümlichen Liede Felix Mendelssohn Bartholdys', *Hans Albrecht in memoriam* (Kassel, 1962), 200

K.-H. Köhler: 'Das Jugendwerk Felix Mendelssohns: die vergessene Kindheitsentwicklung eines Genies', *DJbM*, vii (1962), 18

D. Siebenkäs: 'Zur Vorgeschichte der Lieder ohne Worte von Men-delssohn', *Mf*, xv (1962), 171

E. Werner: 'Mendelssohns Kirchenmusik und ihre Stellung im 19. Jahrhundert', *Kongress Bericht Kassel, 1962*, 207

A. V. Linden: 'A propos du "Lauda Sion" de Mendelssohn', *RBM*, xvii (1963), 124

P. Mies: 'Über die Kirchenmusik und über neu entdeckte Werke bei Felix Mendelssohn Bartholdy', *Musica sacra*, lxxxiii (1963), 212, 246

D. Mintz: 'Mendelssohn's *Elijah* Reconsidered', *Studies in Ro-manticism*, iii (1963), 1

H. Wirth: 'Natur und Märchen in Webers "Oberon", Mendelssohns "Ein Sommernachtstraum" und Nicolais "Die lustigen Weiber von Windsor"', *Festschrift Friedrich Blume zum 70. Geburtstag* (Kassel, 1963), 389

S. Vendrey: *Die Orgelwerke von Felix Mendelssohn Bartholdy* (diss., U. of Vienna, 1965)

J. Werner: *Mendelssohn's 'Elijah': a Historical and Analytical Guide to the Oratorio* (London, 1965)

O. Jonas: 'An Unknown Mendelssohn Work', *American Choral Review*, ix/2 (1967), 16

H. C. Wolff: 'Zur Erstausgabe von Mendelssohns Jugendsinfonien', *DJbM*, xii (1967), 96

M. Wilson: 'A Mendelssohn Manuscript', *Victoria and Albert Museum Bulletin*, iv (1968), 113 [*Hear my prayer*]

Bibliography

G. Friedrich: *Die Fugenkomposition in Mendelssohns Instrumentalwerk* (Bonn, 1969)

M. Weyer: *Die deutsche Orgelsonate von Mendelssohn bis Reger* (Regensburg, 1969)

A. J. Filosa: *The Early Symphonies and Chamber Music of Felix Mendelssohn Bartholdy* (diss., Yale U., 1970)

J. A. McDonald: *The Chamber Music of Felix Mendelssohn-Bartholdy* (diss., Northwestern U., 1970)

R. Sietz: 'Die musikalische Gestaltung der Loreleysage bei Max Bruch, Felix Mendelssohn und Ferdinand Hiller', *Max Bruch-Studien* (Cologne, 1970), 14

R. Gerlach: 'Mendelssohns Kompositionsweise: Vergleich zwischen Skizzen und Letztfassung des Violinkonzerts op.64', *AMw*, xxviii (1971), 119

——: 'Mendelssohns schöpferische Erinnerung der "Jugendzeit": die Beziehungen zwischen dem *Violinkonzert*, op.64, und dem *Oktett für Streicher*, op.20', *Mf*, xxv (1972), 142

J. Horton: *Mendelssohn Chamber Music* (London, 1972)

T. Stoner: *Mendelssohn's Published Songs* (diss., U. of Maryland, 1972)

M. Thomas: *Das Instrumentalwerk Felix Mendelssohn-Bartholdys: eine systematisch-theoretische Untersuchung unter besonderer Berücksichtigung der zeitgenössischen Musiktheorie*, Göttinger musikwissenschaftliche Arbeiten, iv (Kassel, 1972)

O. Bill: 'Unbekannte Mendelssohn-Handschriften in der Hessischen Landes- und Hochschulbibliothek Darmstadt', *Mf*, xxvi (1973), 345

H. Eppstein: 'Zur Entstehungsgeschichte von Mendelssohns Lied ohne Worte op.62, 3', *Mf*, xxvi (1973), 486

F. Reininghaus: 'Zwischen Historismus und Poesie: über die Notwendigkeit umfassender Musikanalyse und ihre Erprobung an Klavierkammermusik von Felix Mendelssohn Bartholdy und Robert Schumann', *Zeitschrift für Musiktheorie*, iv (1973), 22; v (1974), 34

G. Schuhmacher: 'Zwischen Autograph und Erstveröffentlichung', *BMw*, xv (1973), 253 [op.44]

R. Elvers: 'Auf den Spuren der Autographen von Felix Mendelssohn Bartholdy', *Beiträge zur Musikdokumentation: Franz Grasberger zum 60. Geburtstag* (Tutzing, 1975), 83

W. Goldhan: 'Felix Mendelssohn Bartholdys Lieder für gemischten und Männerchor', *BMw*, xvii (1975), 181

F. Reininghaus: 'Studie zur bürgerlichen Musiksprache: Mendelssohns "Lieder ohne Worte" als historisches, ästhetisches und politisches Problem', *Mf*, xxviii (1975), 34

R. Szeskus: ' "Die erste Walpurgisnacht," op.60, von Felix Mendelssohn Bartholdy', *BMw*, xvii (1975), 171

R. Elvers, ed.: *'Nichts ist so schwer gut zu componiren als Strophen': zur Entstehungsgeschichte des Librettos von Felix Mendelssohns Oper 'Die Hochzeit des Camacho'* (Berlin, 1976)

B. W. Pritchard: 'Mendelssohn's Chorale Cantatas: an Appraisal', *MQ*, lxii (1976), 1

H. Kohlhase: 'Studien zur Form in den Streichquartetten von Felix Mendelssohn Bartholdy', *Hamburger Jb für Musikwissenschaft*, ii (1977), 75

D. Seaton: *A Study of a Collection of Mendelssohn's Sketches and other Autograph Material, Deutsche Staatsbibliothek Berlin 'Mus. Ms. Autogr. Mendelssohn 19'* (diss., Columbia U., 1977)

——: 'A Draft for the Exposition of the First Movement of Mendelssohn's "Scotch" Symphony', *JAMS*, xxx (1977), 129

D. L. Butler: 'The Organ Works of Felix Mendelssohn Bartholdy', *The Diapason*, lxix/3, 5, 7 (1978), lxx/12 (1979), lxxi/2 (1980)

G. Feder: 'Zwischen Kirche und Konzertsaal: zu Felix Mendelssohn Bartholdys geistlicher Musik', *Religiöse Musik in nicht-liturgischen Werken von Beethoven bis Reger* (Regensburg, 1978), 97

F. Krummacher: *Mendelssohn–der Komponist: Studien zur Kammermusik für Streicher* (Munich, 1978)

A. Kurzhals-Reuter: *Die Oratorien Felix Mendelssohn Bartholdys: Untersuchungen zur Quellenlage, Entstehung, Gestaltung und Überlieferung* (Tutzing, 1978)

G. Weiss: 'Eine Mozartspur in Felix Mendelssohn Bartholdys Sinfonie A-Dur op.90 ("Italienische")', *Mozart: Klassik für die Gegenwart* (Oldenburg, 1978), 87

R. L. Todd: 'Of Sea Gulls and Counterpoint: the Early Versions of Mendelssohn's *Hebrides* Overture', *19th Century Music*, ii (1978–9), 197

R. M. Longyear: 'Cyclic Form and Tonal Relationships in Mendelssohn's "Scottish" Symphony', *In Theory Only*, iv/7 (1979), 38

R. L. Todd: *The Instrumental Music of Felix Mendelssohn Bartholdy: Selected Studies Based on Primary Sources* (diss., Yale U., 1979)

R. W. Ellison: 'Mendelssohn's "Elijah": Dramatic Climax of a Creative Career', *American Choral Review*, xxii/1 (1980), 3

R. L. Todd: 'An Unfinished Symphony by Mendelssohn', *ML*, lxi (1980), 293

C. Dahlhaus: 'Hoch symbolisch intentoniert: zu Mendelssohns *Erster Walpurgisnacht*', *ÖMz*, xxxvi (1981), 290

R. L. Todd: 'A Sonata by Mendelssohn', *Piano Quarterly*, xxix (1981), 30

E. Kraus: 'Die formale und motivische Einbindung des Choralthemas

Bibliography

in Mendelssohns erster und Rheinbergers dritter und vierter Orgelsonate', *Gedenkschrift Hermann Beck* (Laaber, 1982), 161

D. Seaton: 'The Romantic Mendelssohn: the Composition of *Die erste Walpurgisnacht*', *MQ*, lxviii (1982), 398

R. L. Todd: 'An Unfinished Piano Concerto by Mendelssohn', *MQ*, lxviii (1982), 80

R. Parkins and R. L. Todd: 'Mendelssohn's Fugue in F minor: a Discarded Movement of the First Organ Sonata', *Organ Yearbook*, xiv (1983)

R. L. Todd: *Mendelssohn's Musical Education: A Study and Edition of his Exercises in Composition* (Cambridge, 1983)

——: 'A Passion Cantata by Mendelssohn', *American Choral Review*, xxv/1 (1983), 3

OTHER STUDIES

F. Niecks: 'On Mendelssohn and some of his Contemporary Critics', *MMR*, v (1875), 162

E. Hanslick: 'Zur Erinnerung an Felix Mendelssohn-Bartholdy', *Am Ende des Jahrhunderts* [*Die moderne Oper*, viii] (Berlin, 1899/*R*1971), 409

L. Leven: *Mendelssohn als Lyriker unter besonderer Berücksichtigung seiner Beziehungen zu Ludwig Berger, Bernhard Klein und Adolph Bernhard Marx* (diss., U. of Frankfurt am Main, 1926)

R. Werner: *Felix Mendelssohn Bartholdy als Kirchenmusiker* (diss., U. of Frankfurt am Main, 1930)

G. Kinsky: 'Was Mendelssohn Indebted to Weber?', *MQ*, xix (1933), 178

L. H. and H. Tischler: 'Mendelssohn's Style: *the Songs without words*', *MR*, viii (1947), 256

M. Kingdon-Ward: 'Mendelssohn and the Clarinet', *MMR*, lxxxiii (1953), 60

R. Sterndale Bennett: 'Mendelssohn as Editor of Handel', *MMR*, lxxxvi (1956), 83

H. C. Wolff: 'Mendelssohn and Handel', *MQ*, xlv (1959), 175

D. Mintz: 'Mendelssohn's Water Color of the *Gewandhaus*', *Notes*, xviii (1960–61), 211

W. Vetter: 'Die geistige Welt Mendelssohns', *Vermächtnis und Verpflichtung: Festschrift für Franz Konwitschny* (Leipzig, 1961), 38

N. Temperley: 'Mendelssohn's Influence on English Music', *ML*, xliii (1962), 224

H. Engel: 'Die Grenzen der romantischen Epoche und der Fall Mendelssohn', *Festschrift O. E. Deutsch* (Kassel, 1963), 259

D. Mintz: 'Mendelssohn and Romanticism', *Studies in Romanticism*, iii (1964), 216

M. Geck: *Die Wiederentdeckung der Matthäuspassion im 19. Jahrhundert* (Regensburg, 1967)

J. A. Mussulman: 'Mendelssohnism in America', *MQ*, liii (1967), 335

H. C. Wolff: 'Das Mendelssohn-Bild in Vergangenheit und Gegenwart', *Musa-Mens-Musici, im Gedenken an W. Vetter* (Leipzig, 1967), 321

S. Grossmann-Vendrey: 'Mendelssohn und die Vergangenheit', *Die Ausbreitung des Historismus über die Musik* (Regensburg, 1969), 73

G. Schuhmacher: 'Mendelssohn heute oder seine Wiederentdeckung durch die Schallplatte', *Schallplatte und Kirche*, i (1969), 3

R. Elvers: 'Felix Mendelssohns Beethoven-Autographe', *Kongress Bericht Bonn 1970*, 380

F. Krummacher: 'Über Autographe Mendelssohns und seine Kompositionsweise', *Kongress Bericht Bonn 1970*, 482

E. Glusman: 'Taubert and Mendelssohn: Opposing Attitudes toward Poetry and Music', *MQ*, lvii (1971), 628

G. Hendrie: *Mendelssohn's Rediscovery of Bach* (London, 1971)

J. Forner: 'Mendelssohns Mitstreiter am Leipziger Konservatorium', *BMw*, xiv (1972), 185

——: 'Mendelssohn und die Bachpflege in Leipzig', *Arbeitsbericht zur Geschichte der Stadt Leipzig*, ii/23 (1972), 85

K.-H. Köhler: 'Das dramatische Jugendwerk Felix Mendelssohn Bartholdys–Basis seiner Stil- und Persönlichkeitsentwicklung', *Kongress Bericht Copenhagen 1972*, 495

L. Lockwood: 'Mendelssohn's Mozart: a New Acquisition', *Princeton University Library Chronicle*, xxxiv/1 (1972), 62

C. Lowenthal-Hensel, ed.: *Beiträge zur neueren deutschen Kultur- und Wirtschaftsgeschichte: Mendelssohn-Studien*, i–v (Berlin, 1972–82)

E. Rudolph: 'Mendelssohns Beziehungen zu Berlin', *BMw*, xiv (1972), 205

G. Schönfelder: 'Zur Frage des Realismus bei Mendelssohn', *BMw*, xiv (1972), 169

R. Meloncelli: 'Felix Mendelssohn-Bartholdy, proposta per una nuova prospettiva critica', *Scritti in onore di Luigi Ronga* (Milan, 1973), 331

M. Wehnert: 'Mendelssohns Traditionsbewusstsein und dessen Widerschein im Werk', *DJbM*, xvi (1973), 5

E. Werner: 'Mendelssohn–Wagner: eine alte Kontroverse in neuer Sicht', *Musicae scientiae collectanea: Festschrift Karl Gustav Fellerer* (Cologne, 1973), 640

C. Dahlhaus, ed.: *Das Problem Mendelssohn* (Regensburg, 1974)

Bibliography

J. Godwin: 'Early Mendelssohn and Late Beethoven', *ML*, lv (1974), 272

O. Wessely: 'Bruckners Mendelssohn-Kenntnis', *Bruckner-Studien* (Vienna, 1975), 81

L. Nowak: 'Mendelssohns "Paulus" und Anton Bruckner', *ÖMz*, xxxi (1976), 574

T. Stoner: 'Mendelssohn's Lieder not included in the Werke', *FAM*, xxvi (1979), 258

R. Elvers: 'Verlorengegangene Selbstverständlichkeiten: zum Mendelssohn-Artikel in *The New Grove*', *Festschrift Heinz Becker zum 60. Geburtstag* (Laaber, 1982), 417

F. Krautwurst: 'Felix Mendelssohn Bartholdy als Bratschist', *Gedenkschrift Hermann Beck* (Laaber, 1982), 151

G. Schuhmacher, ed.: *Felix Mendelssohn Bartholdy* (Darmstadt, 1982)

E. Werner: 'Felix Mendelssohn–Gustav Mahler: Two Borderline Cases of German-Jewish Assimilation', *Yuval*, iv (1982), 240

T. Ehrle: *Die Instrumentation in den Symphonien und Ouvertüren von Felix Mendelssohn Bartholdy* (Wiesbaden, 1983)

N. Thistlethwaite: 'Bach, Mendelssohn, and the English Organist: 1810–1845', *British Institute of Organ Studies*, vii (1983), 34

J. W. Finson and R. L. Todd, eds.: *Mendelssohn and Schumann: Essays on their Music and its Context* (Durham, 1984)

R. Parkins: 'Mendelssohn and the Erard Piano', *Piano Quarterly*, xxxii (1984), 53

Index

Aachen: Lower Rhine Music Festival, 231
Abbotsford, Scotland, 209
Abruzzi mountains, 102, 140
Acts of the Apostles, 214–5
Albert, Prince Consort, 223, *224*
Alexandre, Edouard, 124
Allgemeine musikalische Zeitung, 2
All people that on earth do dwell (*Old 100th*), 118
Alps, 87, 204
Aquinas, Thomas: *Lauda Sion Salvatorem*, 231
Argand, Aimé, 37
Arnim, Archim von: *Des Knaben Wunderhorn*, 55
Arnim, Bettina von: *Dies Jahr gehört dem König*, 223
Attwood, Thomas, 211
Auber, Daniel-François-Esprit, 25
——, *Fra Diavolo*, 214
Augsburg, 2, 3
Augsburg Confession, 254
Austria, 38, 110, 112

Bach, Carl Philipp Emanuel, 63
Bach, Johann Sebastian, 131, 203, 208, 211, 217, 229, 235, 246, 253, 254, 267
——, Cantatas, 246, 250
——, Concerto in D minor BWV1063, 217
——, Memorial at Leipzig, 220–21
——, Orchestral suites, 217
——, Passions, 246
——, *St Matthew Passion*, 146, 208–9, *210*, 212
——, *Wachet auf, ruft uns die Stimme*, 246

Bad Doberan, nr. Rostock, 205, 259
Baden, nr. Vienna, 22
Baden-Baden, 119, 123, 233
Baedeker, Karl, 31
Baermann, Heinrich, 7, 8, 59, 61
Baillot, Pierre, 205, 212
Baini, Giuseppe, 211
Balakirev, Mily Alexeyevich, 168
Balfe, Michael William: *The Maid of Honour*, 116
Ballanche, Pierre-Simon: *Erigone*, 137
Balzac, Honoré de, 97
Bamberg, 7
Barbaia, Domenico, 19
Barbier, Auguste, 106
Barzun, Jacques, 169
Bartholdy, 201; *see also* Mendelssohn(-Bartholdy) family members
Basedow, Johann Bernhard, 238
Basle, 31
Becker, Carl Ferdinand, 226
Beer, Jacob [Jakob]: *see* Meyerbeer, Giacomo
Beethoven, Ludwig van, 22, 30, 38, 96, 103, 105, 112, 134, 138, 139, 140, 149, 155, 164, 165, 167, 201, 204, 212, 217, 229, 235, 260, 264
——, *An die ferne Geliebte*, 57
——, *Coriolanus* Overture, 230
——, *Fidelio*, 12, 22, 39, 237
——, *Leonore* Overture no.1, 216
——, *Leonore* Overture no.3, 213
——, Mass in C, 220
——, Piano Concerto no.5, 209, 223
——, Piano Trios op.1, 264
——, String Quartet in E♭ op.74, 264
——, Symphony no.3, 96
——, Symphony no.4, 215

303

Index

305

Index

307

Index

Index

Paer, Ferdinando, 25, 38
Paganini, Nicolò, 105, 107
Palestrina, Giovanni Pierluigi da, 208, 211
Paris, 25, 28, 66, 89, 90, 91, 92, 95, 98, 99, 102, 103, 104, 109, 110, 111, 113, 114, 116, 117, 118, 119, 120, 127, 134, 135, 199, 201, 205, 212, 245
——, Conservatoire, 91, 92, 96, 104, 114, 116, 121, 122, 124, 212
——, Prix de Rome, 92–3, 98, 151
——, Ecole de Médecine, 89
——, Exposition Universelle, 120, 122
——, Grand Festival de l'Industrie, 112
——, the Invalides, 107, 145
——, Montmartre, 104
——, Montmartre Cimetière, 127
——, Odéon theatre, 93, 141
——, Opéra, 90, 106, 109, 110, 122, 123, 135, 137, 165
——, Opéra-Comique, 114, 135
——, Palais de l'Industrie, 120
——, Revolution of 1830, 107
——, St Eustache, 119, 147
——, St Roch, 91, 98
——, Société des Concerts, 96, 119, 120, 164
——, Société Philharmonique, 119
——, Théâtre des Nouveautés, 92
——, Théâtre Franconi, 112
——, Théâtre-Lyrique, 123
Pasdeloup, Jules Etienne, 124
Passau, 114
Paton, Mary Anne, 28
Pest, 113, 114
Pestalozzi, Johann Heinrich, 238
Planché, James Robinson, 24, 26, 50, 239
Pleyel, Camille, 102
Pleyel, Ignace Joseph: Quartets, 89
Pleyel, Marie: see Moke, Camille
Poissl, Johann Nepomuk, 40
——, *Athalia*, 31

Pokój: *see* Carlsruhe
Pompeii, 103, 212
Potsdam, 204, 222
——, Neuer Palais, 222, 227, 229
Prague, 4, 8, 10, 11, 12, 13, 14, 16, 21, 23, 30, 31, 35, 36, 38, 61, 113, 114, 200
——, Opera, 10, 12, 14, 16
Pressburg (now Bratislava), 211
Prod'homme, J.-G., 169
Prussia, 199, 222

Quedlinburg, 23

Racine, Jean: *Athalie*, 229
Rakemann, Louis, 217
Rameau, Jean-Philippe, 131, 138
——, *Traité de l'harmonie*, 88
Ramsgate, 231
Rauch [horn player], 13
Ravel, Maurice, 160, 168
Recio, Marie(-Geneviève), 111, 121; *see also* Berlioz, Marie
Reger, Max, 269
Reicha, Antoine, 92, 156
Reichardt, Johann Friedrich, 252
Rellstab, (Heinrich Friedrich) Ludwig, 204, 249
Revue musicale, 103
Reyer, (Louis-Etienne-)Ernest, 127, 168
Rhode, Johann Gottlieb, 3, 4
Richter, J. P. F. [Jean Paul], 21, 207
Riemann, Hugo, 269
Rietz, Eduard, 205, 263
Rietz, Julius, 218, 223, 233, 234, 268
Righini, Vincenzo, 217
Rimsky-Korsakov, Nikolay Andreyevich: *Antar*, 168
Ritter, Carl, 208
Robert, Alphonse [Berlioz's cousin], 90
Rochlitz, (Johann) Friedrich, 2, 5
Rode, Pierre, 205
Romberg, Andreas, 12

311

Index

313